WHAT ARE

Gamma-Ray Bursts?

JOSHUA S. BLOOM

PRINCETON UNIVERSITY PRESS

PRINCETON AND OXFORD

Copyright © 2011 by Princeton University Press

Published by Princeton University Press, 41 William Street,
Princeton, New Jersey 08540
In the United Kingdom: Princeton University Press, 6 Oxford Street,
Woodstock, Oxfordshire OX20 1TW
press.princeton.edu

Library of Congress Cataloging-in-Publication Data

Bloom, Joshua S., 1974–
What are gamma-ray bursts? / Joshua S. Bloom.
        p. cm. – (Princeton Frontiers in physics)
Includes bibliographical references and index.
ISBN 978-0-691-14556-3 (hardcover : alk. paper) –
ISBN 978-0-691-14557-0 (pbk. : alk. paper)
1. Gamma ray bursts.   2. Stars–Formation.   I. Title.
QB471.7.B85B56 2011
523.01′97222–dc22   2010038975

British Library Cataloging-in-Publication Data is available

This book has been composed in Garamond
Printed on acid-free paper ∞
press.princeton.edu

Typeset by S R Nova Pvt Ltd, Bangalore, India
Printed in the United States of America

10 9 8 7 6 5 4 3 2 1

To Sofia and Claudia,
who tried their darndest
to make sure that this book never got written.

And to Anna,
who did her best to make sure that it did.

CONTENTS

PREFACE

In a strictly observational sense the question "What are gamma-ray bursts?" has a straightforward answer: gamma-ray bursts (GRBs) are unannounced flashes of high-energy light detected from seemingly random places on the sky. In the first few seconds of activity, GRBs emit as much energy as our Sun will release in its entire ten billion-year lifetime. We now believe that they are the brightest *electromagnetic events* in the Universe. The arrival at this mind-boggling conclusion is the result of a remarkable journey spanning the last three decades of the twentieth century. This historical tale—of an unexpected beginning, mystery, cutting-edge technology, and persistent scientists—is the main subject of chapter 1 and frames the content of the entire book.

"What *are* gamma-ray bursts?"—as in "what are the astrophysical origins of gamma-ray bursts?" and "what are the uses of gamma-ray bursts in helping us to understand the universe?"—is a deeper question with less certain answers. As discussed in chapter 2, the event themselves are very likely due to various interactions of particles and light with magnetic fields. These basic ingredients are carried along by an explosion of material moving at speeds close to the speed of light (chapters 3 and 4). The object or objects responsible for this fantastic explosion are, as we shall see in chapter 5, of great interest to astronomers. There is strong evidence that some GRBs are produced when stars much more massive than the Sun run out of nuclear fuel and explode. Some GRBs are likely due to gigantic magnetic flares near the surface of exotic dead stars called *neutron stars*. The quest to understand the diversity of the astrophysical entities that make GRBs is an ongoing one and represents one of the most pressing inquiries in modern astrophysics.

By virtue of their special properties, GRBs are emerging as unique tools in the study of broad areas of astronomy and physics (chapter 6). Indeed, while GRB astronomers continue to broaden the understanding of their nature and origin, others are using GRBs to shed light (literally) on some of the most basic questions of our time: How and when do stars form? How do the most massive stars end their lives? What makes black holes? What is the age of the Universe and what is its ultimate fate? Is very fabric of space and time different from what Einstein imagined? In this sense, gamma-ray bursts are not just one of nature's most fantastic firecrackers but unique and powerful tools in the study of the Universe itself.

At the core, this book is meant as a thorough introduction to the relevant technical and physical concepts of the field. There are more than ten thousand papers published on the subject of GRBs,[1] and so a distillation of the main concepts and observations is a particularly important. This is a book intended for astronomy enthusiasts and those with some level of training in undergraduate physical sciences. However, I presuppose no particular or specialized knowledge of astronomy and physics here. Important technical topics, concepts, and useful jargon (highlighted first in italics) are either defined in the text or in the glossary toward the back of the book. Less important phrases or concepts are given in quotes. References to seminal discovery papers and other primary sources are provided in the notes. Equations are used (but sparingly). I present graphs and figures, most adapted directly from academic research papers; the adage that a "picture is worth a thousand words" holds for astrophysicists too!

This book is also intended as a point of departure for those interested in pursuing the subject matter in more depth: suggested readings (both popular and academic in nature) are highlighted at the back of the book. Someone with a strong physics and astronomy training, looking for more depth than what is provided here, will find a relatively recent comprehensive review of both theory and phenomenology in *Gamma-Ray Bursts: The Brightest Explosions in the Universe*, by Gilbert Vedrenne and Jean-Luc Atteia (2009).

I am indebted to all those who have endeavored to push to the technical and scientific forefront of the field. I thank my research advisors: Edward E. Fenimore, Jonathan (Josh) E. Grindlay, Martin J. Rees, Ralph A.M.J. Wijers, Dale A. Frail, and Shrinivas R. Kulkarni. As scientists they each contributed in remarkable and unique ways to the development of the GRB field; as mentors they had their own profound impact on my professional development. I am particularly grateful for the years of interactions with my GRB-oriented graduate students (Andrew Friedman, Adam Morgan, and Daniel Perley) and postdoctoral scholars (Nathaniel Butler, S. Bradley Cenko, Bethany Cobb, Daniel Kocevski, Maryam Modjaz, and David Pooley). I am extraordinarily thankful for years of collaborations with many of the people whose work is presented here, especially J. Xavier Prochaska and Hsiao-Wen Chen. This book benefitted enormously from the feedback of anonymous reviewers on a draft of the manuscript as well as close reads from Mark Galassi, Adam Morgan, Bethany Cobb, Maryam Modjaz, S. Bradley Cenko, and Daniel Perley. All errors and omissions that remain in the published book are entirely my own. Enjoy!

Joshua S. Bloom
Berkeley, California
July 2010

WHAT ARE

Gamma-Ray Bursts?

1

INTRODUCTION

Serendipity is jumping into a haystack to search for a needle, and coming up with the farmer's daughter.

—Julius H. Comroe Jr.

1.1 Serendipity during the Cold War

Before *Mythbusters* and *The A-Team* made big explosions cool, big explosions were decidedly *uncool*. The threat of nuclear war between the United States and the USSR (and, perhaps, China)—made blatantly real during the Cuban Missile Crisis in October 1962—had become a fixture in everyday life. One year after the crisis, seeking to diffuse an escalating arms race and the global increase of radioactive fallout from nuclear weapons testing, Soviet Premier Nikita Khrushchev and U.S. President John F. Kennedy agreed to the Partial Test Ban Treaty. Ratifying nations agreed that all nuclear weapons testing would be conducted underground from then on: no longer would tests be conducted in oceans, in the atmosphere, or in space.

The United States, led by a team at the Los Alamos National Laboratory, promptly began an ambitious space satellite program to test for "non-compliance" with the Partial Test Ban Treaty. The existence of the Vela[1] Satellite

Program was unclassified: the rationale, experimental design, and satellite instrumentation were masterfully detailed in peer-reviewed public journals while the program was on going.[2] The concept for this space-based vigilance endeavor was informed by the physics of nuclear explosions: while the optical flash of a nuclear detonation could be shielded, the *X-rays*, *gamma rays* (sometimes written as γ-rays), and *neutrons* that are produced in copious numbers in the first second of an explosion are much more difficult to hide; we call the measurement of these by-products the "signature" of a nuclear detonation. Going into space for such surveillance was a must: the Earth's atmosphere essentially blocks X-rays, gamma rays, and neutrons from space.

While the signatures of nuclear detonations were well understood, the *background radiation* of light and particles in space was not. To avoid false alarms caused by unknown transient enhancements in the background, satellites were launched in pairs—both satellites would have to see the same very specific signatures in their respective instruments for the alarms bells to sound. Widely separated satellite pairs also had the advantage that most of the Earth could be seen at all times. While the Vela orbits provided little vantage point on the dark side of the Moon—a natural location to test out of sight—the gamma rays and neutrons from the expanding plume of nuclear-fission products would eventually come into view. In total, six pairs (Vela 1a,b through Vela 6a,b) were launched between 1964 and 1970.

As evidenced by the Vela Satellite Program, the U.S. was obviously very serious about ensuring compliance.

That the capabilities of the program were open was also a wonderful exercise in cold war gamesmanship—you are much less inclined to break the rules if you are convinced you will get caught.

While hundreds of thousands of events were detected by the Velas—mostly from lightning on Earth and charged particles (*cosmic rays*) hitting the instruments—the telltale signatures of nuclear detonation were thankfully never discovered.[3] Those events that were obviously not of pernicious or known origin were squirreled away for future scrutiny.[4]

Starting in 1969, Los Alamos employee Ray Klebesadel began the laborious task of searching, by eye, the Vela data for coincident gamma-ray detections in multiple satellites. One event, from July 2, 1967,* stood out (figure 1.1). Seen in both the gamma-ray detectors of Vela 4a and Vela 4b (and weakly in the less sensitive Vela 3a and Vela 3b detectors), the event was unlike any known source. Though there was no known solar activity on that day, the event data themselves in one satellite were incapable of ruling out a Solar origin, especially if it was a new sort of phenomenon from the Sun. Over the next several years, other intriguing events similar to the July 2nd event were seen in the Vela data. By 1972, Klebesadel and

---

*GRBs are conventionally identified by the date that they are detected. The first two numbers correspond to the year, the next two correspond to the month, and the last two correspond to the day. In the case that more than one GRB is detected on the same day, letters are appended to the burst name. For instance, the first GRB detected was GRB 670702. GRB 101221B would be the second GRB detected on December 21, 2010. (This naming convention clearly runs into problems after a hundred years of observations!)

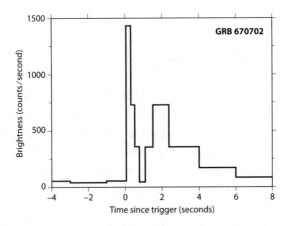

Figure 1.1. The first gamma-ray burst, GRB 670702, detected by the Vela 3a,b and 4a,b satellites. Shown is the gamma-ray light curve of the event, which is the instrumental brightness (counts per second) versus time measured since the event triggered the on-board instruments. The count rate before the burst is not zero due to persistent gamma-ray sources in the sky and random instrumental events in the detector. But when the event arrives at the detector, it vastly outshines the background. GRB670702 was a long-duration GRB, lasting more than eight seconds and showing variability on *timescales* less than one second. Adapted from J. Bonnell, *A Brief History of the Discovery of Cosmic Gamma-Ray Bursts.* http://antwrp. gsfc.nasa.gov/htmltest/jbonnell/www/grbhist.html (1995).

his colleagues Ian Strong and Roy Olson had uncovered sixteen such events using automated computer codes to aid with the arduous searches.

What were these bursts of gamma rays? To answer that question, the Los Alamos team recognized that it had better determine where on the sky the events came from. Pinpointing the direction of a light source is easy if you can focus it: this is what cameras used for photography and the human eye do well with visible light. But X-rays, and especially gamma rays, are not amenable to focusing: the energies of these *photons* are so high that they do not readily interact with the free electrons in metals and so cannot be reflected to large angles. The focusing of light without large-angle reflection is exceedingly difficult. The best the X-ray and gamma-ray detectors on the Velas could do was *stop* those photons, recording both the energy deposited in the detectors and the time that the photon arrived at the satellite.

The arrival time of the photons from specific events held the key to localization. Just as a thunderclap is heard first by those closest to the lightening bolt, an impulsive source of photons would be seen first in the satellite closest to the event and then later, after the light sweeps by, with the more distant satellite. Light (and sound, in the case of thunder) has a finite travel speed. Since the Vela satellites were dispersed at large distances from each other (approximately 200,000 kilometers) the difference in the arrival times of the pulses could be used to reconstruct the origin on the sky, the location on the *celestial sphere*. As figure 1.2 shows, an event seen in two satellites produces an annular location on the sky, and an event seen in three satellites produces a location in two patches on the sky.*

---

*It turns out that, with precise-enough timing, a network of satellites can localize sources within the Solar System not only on the sky (two dimensions) but also in the third dimension (distance). A few GRBs have been shown to be unequivocally at distances beyond the Solar System using such a timing technique.

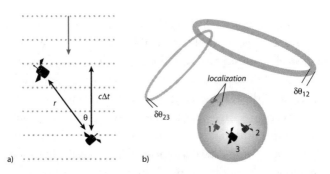

Figure 1.2. Triangulation of gamma-ray bursts using the arrival time of light at different satellites. Image (a) shows the geometry of localization in two dimensions. The GRB propagates down through the Solar System as a *plane wave* depicted here with dashed lines. The event triggers the satellite at left first, then the satellite at right. By measuring the difference in trigger time (Δt) and knowing the distance r between the satellites, the angle θ toward the event can be inferred. Here c is the speed of light. Image (b) shows the localization on the *celestial sphere* (gray circle) using three satellites. Two independent angles are determined using the difference in arrival times between satellites 1–2 and 2–3. The uncertainty in those angles $\delta\theta$ is directly determined by the uncertainty in the precise time difference between satellite triggers. The direction toward the GRB is determined by the regions on the sky where the annuli overlap.

This triangulation capability, albeit crude, was sufficient to convince the Los Alamos team that it had uncovered a class of events that was not coming from the Earth, Sun, Moon, or any other known Solar System object. In 1973, Klebesadel, Strong, and Olson published their findings in the *Astrophysical Journal,* one of the venerable peer-reviewed journals used for describing scientific results in astronomy. The paper[5] titled "Observations of Gamma-Ray Bursts of Cosmic Origin" marked the beginning of the gamma-ray burst (GRB)[6] enigma that to this day captivates the imagination and keeps astronomers scratching their heads.

The word serendipity is overused and misused in science. Most mistake a serendipitous discovery to be synonymous with an unexpected (and unforeseen) discovery. But, as Julius Comroe's colorful analogy in the epigraph describes, serendipity demands both an unexpected discovery and an entirely more pleasant discovery than the one being pursued. While GRBs certainly were unexpected and unforeseen,[7] they were also much more scientifically valuable than what was being sought after: instead of the detection of a nuclear test by an enemy, a discovery that in the 1960s would have set the world down a dangerous and dark path, GRBs were a fresh light from the dark heavens. Indeed, their mysterious nature would captivate a generation of astronomers. The discovery of GRBs—not just their detection but the recognition that the events represented a new phenomenon in nature—was truly a serendipitous moment in modern science.

1.2 A New Field Begins

Members of Klebesadel's team announced the discovery of GRBs at the June 1973 meeting of the American Astronomical Society, a few days after the publication of their seminal paper. In that meeting (and in the paper) they described their observations testing the hypothesis that

GRBs originated from *supernovae* (SNe) in other galaxies; this was the only physical model for the origin of cosmic bursts of gamma rays available at the time.[8] By trying to correlate a GRB in time and sky position to all known SNe, the attempt to connect GRBs to the then-brightest explosions in the universe "proved uniformly fruitless."[9]

Determining what objects and what events on those objects produced GRBs quickly became a hot topic. By the end of 1974, more than one dozen ideas for the origin of GRBs had already been published. The theories spanned an astonishing range of possibilities, from sunlight scattering off fast-moving dust grains to comets colliding with *white dwarfs* (WDs) to "antimatter asteroids" smashing into distant stars. All viable models necessarily accommodated the available data, but the GRB data were simply too sparse to constrain a talented and imaginative group of eager scientists.

More data would be needed to narrow down the range of plausible models. By the end of 1973, the Los Alamos team had found a total of twenty-three GRBs.[10] Teams working with other satellites equipped with gamma-ray detectors also began reporting detections of GRBs,[11] even some of the same events seen by the Vela satellites. New programs were conceived to find more GRBs and observe them with more sensitive detectors. The supposition—if not just a hope—was that with better data some telltale signature of the origin of the events would emerge. Unbeknown to those sprinting to find the answer, for all but a few special events, those telltale signatures would take over thirty years to uncover (a veritable marathon in modern science).

Light does not easily betray its origin: there is nothing in a gamma-ray photon itself that can tell us how far it traveled, nor can we learn directly just how many of those energetic photons streamed away from the event that produced the GRB. Without a measurement of the distance to a source, the pool of possible culprits is simply too broad: since we have a general sense of the types and the spatial distribution of objects in a given volume of space, if we knew that GRBs arose from distances on the size scale of the Solar System (for example), then there could be only a select set of objects responsible (comets, asteroids, planets, etc.). At a more fundamental level, without knowledge of distance, it is all but impossible to know how much energy the source put out. And without that knowledge the range of physical mechanisms that could be responsible for the sudden release of all that energy is also too broad. Case in point: a street lamp appears about as bright as the Sun, yet the scales of energy output are vastly different as are the physical origins of the light.

Since light does not directly encode distance, how do astronomers determine distance to astrophysical entities? If sufficiently nearby, objects appear to be in slightly different places on the sky for observers at different places. This measurement of *parallax* yields a direct triangulation of distance but is exceedingly difficult to determine for most objects beyond a few hundred *light years* away from Earth. Beyond that, for all but a few special cases, we must infer distance by associating some source with a source whose intrinsic brightness or size we think we know (usually because we think there is an analogous system within the parallax volume).

The key, then, for GRBs would be to associate the events with something else whose distance we could more readily infer. In this respect, the inability to measure a precise two-dimensional position of a GRB on the sky directly hampered the ability to measure the all-important third dimension. Getting better positions of GRBs on the sky became the driving impetus behind the next several decades of GRB observational projects.

1.3 Precise Localizations and the Search for Counterparts

By the late 1970s, not only were there more satellites flying with higher-sensitivity detectors, but some of these satellites were far from Earth (in particular, near Venus and the Sun). This *interplanetary network* (IPN) gave a significant improvement on the timing localizations of GRBs (see figure 1.2). At a distance of up to $d = 2$ *astronomical units* (AU) (twice the distance from the Earth to the Sun), a pair of satellites with the capability to determine the time of the onset of a GRB to an accuracy of $\delta t = 0.1$ seconds would be able to produce an annular localization ring of thickness $\delta\theta \approx \delta t \times c/d = 10^{-4}$ *radian* $\approx 1/3$ *arcminute*. By 1980, there were a handful of well-localized (to tens of square arcminutes or better[12]) GRBs,[13] and by the end of the 1980s there were dozens of well-localized GRBs using the interplanetary timing technique.

In a spatial area on the sky, while millions of times more accurate than the first GRB positions, these square-arcminute localizations proved insufficient to rule out most models. If all *error boxes* on the sky contained a bright star or a bright galaxy, the association with a certain physical class of objects would be secure. This was not the case. Instead, GRBs must have been associated with something faint or unseen. The enormity of the Universe and its bountiful constituents is a real shackle in this respect: in even the most empty directions looking out through our Galaxy, a single error box would contain tens of thousands of faint stars and tens of thousands of faint and distant galaxies. This amounted to a line up of culprits simply too big to get any significant traction on the question of distance and, ultimately, the origin of GRBs.

Observing at gamma-ray wavelengths is just about the worst idea if the goal is to localize an event precisely. But if a *counterpart* at some other wavelength could be associated positively with a specific GRB, then the location of the GRB could be more precisely identified. The most credible counterpart would be an event, consistent with the GRB position, that seemed to happen at around the same time as the GRB—it is actually quite natural to expect that some energy should be pumped into channels other than gamma rays, but just how much energy and on what *timescales* that energy would emerge across the *electromagnetic spectrum* were not well known. As mentioned, no (visible-light) supernova counterparts were found by Klebesadel's team during the early years of the field. And, despite several efforts in the 1970s and 1980s to discover a concurrent signal from radio to infrared to optical to X-ray wavelengths, no convincing counterparts were found. There was another possibility: if the "engine" (see §2.3) that produced the GRB had been active previously, then perhaps a transient counterpart could be found in the old image archives of the same place on the sky. Some tantalizing archival transients were indeed uncovered, but none proved robust under detailed scrutiny.

1.4 The March 5th Event and Soft Gamma-Ray Repeaters

On March 5, 1979, an intense gamma-ray event triggered the IPN satellites distributed throughout the inner Solar System. Within the first tens of milliseconds, the event became so intensely bright that the detectors on board all nine satellites—even those pointing away from the event direction—saturated: photons arrived at such an appreciable rate that they could not be recorded fast enough. This blinding was only temporary, however, as for the next few minutes some detectors recorded a fading signal with an unusual character. Unlike all the other GRBs that had been seen to date, this decaying tail appeared to vary *periodically*. The fact that the initial pulse "turned on" so rapidly suggested that the size of the emitting surface was small, less than the size of the Earth.[14] The eight-second periodicity in the signal was also an important clue for understanding the progenitor. In nature there are only a few classes of physical configurations that give rise to periodic brightness changes; of the most interest are the pulsations of an emitting surface, oscillations through an emitting object, and rotation. The natural (most physically simple) timescale for changes in pulsations and oscillations is the time τ it takes for sound waves to cross the object, $\tau \approx l/c_s$ (where l is the characteristic size of the object and c_s is the speed of sound in the object). For rotation, that timescale is the period of the rotating object. Ordinary stars, like the Sun, have much longer sound-crossing times and rotation periods than eight seconds. On timing arguments alone, one is quickly pushed to consider a very dense (and hence large c_s) and/or small object as the likely origin of such an event.

The March 5th event, by virtue of its very short and intense pulse, was a superb client for timing localization by the IPN. Initial analyses led the event to be localized to a few square arcminutes, which was further refined to 0.1 square arcminutes with multiply redundant consistency checks, given the number of satellites participating in the timing network. Remarkably, the event appeared to lie on the outskirts of a well-known *supernova remnant*[15] in the very nearby galaxy called the *Large Magellanic Cloud* (LMC).[16]

While the distance to the event could not be unambiguously determined, the evidence connecting it to something small/dense and the SN remnant was overwhelming. It became readily apparent that a *neutron star* (NS)* very likely was the cause of the March 5th event. First, NSs are small and dense enough to account for the rapid rise-time observation. Second, NSs were known to be rapidly spinning, some now observed as fast as five-hundred times

---

*A neutron star (NS) is an inert mass supported by *degeneracy pressure* from neutrons and nuclear forces. It is a close analog of WDs (supported by electron degeneracy) but at higher mass and density.

per second (to be sure, the eight-second periodicity was actually longer than most known NS rotation periods). Third, rotating NSs were known to exhibit pulsed emission, though usually at radio wavelengths and not at gamma-ray energies. Fourth, NSs are the by-product of most types of supernovae, receiving substantial "kicks" of up to ~1,000 km/s during their formation. This natal kick provided a natural explanation for why the event would be located off-center from the SN remnant—in the time since the SN (reckoned to be about 5,000 years from analysis of the SN remnant), the neutron star would have traveled an appreciable distance from the center of the explosion.

If the March 5th event occurred in the LMC, about $r_{\text{LMC}} = 75$ kiloparsec (*kpc*) from the Sun, the total energy release ($E_{\text{March 5th}}$) could now be easily calculated using the $1/r^2$ law:

$$E_{\text{March 5th}} = 4\pi r_{\text{LMC}}^2 S \approx 7 \times 10^{43} \text{ erg,}$$

where $S \approx 10^{-4}$ erg cm^{-2} was the total *fluence* of the event recorded.* This $E_{\text{March 5th}}$ is a fantastically high energy release for an event dominated by a giant subsecond pulse of gamma rays. This amounts to as much energy as the Sun emits in more than a thousand years. Still, this is a small fraction of the *kinetic energy*[17] available to the NS and an even smaller fraction of the available *restmass energy* (see §2.2).

---

*When discussing physical quantities, I generally use the conventional centimeter-gram-second system (cgs) beloved by astronomers. Units of energy are often given as *erg* (from the Greek word, *ergon*, meaning "work"). One erg is approximately the *kinetic energy* of a mosquito buzzing around at ~30 cm/second. One erg equals 10^{-7} Joules.

While the detailed physics of the mechanism(s) at play would be debated for years, the inference of the March 5th event as having arisen from a neutron star was a remarkable feat.* Nevertheless, the March 5th event was quickly realized as anomalous when compared to the other dozens of GRBs observed previously. Aside from the periodicity and the localization in a satellite galaxy of the Milky Way, (a) it was hundreds of times brighter than the brightest events seen in the entire decade of GRB observations, (b) the initial pulse was shorter in duration than in all but $\sim$5 percent of known events, (c) the pulsating tail was "softer"—having a lower characteristic frequency where most of its energy was radiated—than other events. Perhaps the most unusual observation was that there were more events from the same location on the sky, the first just fourteen hours after the March 5th event, then more in the following month. Repetition of other GRBs had never been witnessed previously. The inferred energy release was difficult to reconcile with the demographics of the GRB population as a whole: if all GRBs were variants on the March 5th event, then either March 5th was too bright (if all GRBs were of Galactic origin) or too faint (if all GRBs were of extragalactic origin).

Rather than claim victory in finally understanding the origin of GRBs, those that wrote about the March 5th event immediately touted its oddities compared to the rest

---

*More than one decade later, the quiescent X-ray emission from the location of the March 5th event was found, confirming the basic expectations of a hot NS progenitor.

of the population. Indeed, while some events have been even more energetic than that of March 5th, the event of March 5th is now considered a prototype of a giant flare from a special subclass of GRBs called *Soft Gamma-ray Repeaters* (SGRs). Over the past three decades, six more SGRs were discovered and another two had similar characteristics to the March 5th event without confirmed repetition.[18] All these events appear to have occurred from neutron stars within the Milky Way. In 1998, the spin of another SGR was observed[19] to slow at a rate consistent with braking due to strong magnetic fields. It is now believed that all SGRs in our Galaxy (and a class of anomalous pulsars that radiate at X-ray *wavebands*) are so-called *magnetars*, NSs with extraordinarily high magnetic fields.[20]

1.5 BATSE and the Great Debate

Despite marked advances in understanding the origin, distance scale, and energetics of the March 5th SGR, the mystery for the majority of events raged. Throughout the 1980s, data collection on new GRBs proceeded apace and, by the end of that decade, hundreds of events had been reasonably well localized on the sky and their basic properties measured. Two distinct global properties of "classical GRBs" began to emerge—the location and the brightness distributions—both with important implications for the distance scale of GRBs and hence their origin.

The locations of GRBs on the sky appeared to be randomly (*isotropically*) distributed: that is, there was no indication that any one direction on the sky was especially more apt to produce GRBs than any other (see figure 1.3). If GRBs were due to neutron stars strewn throughout the disk of the Galaxy, for instance, the locations of GRBs on the sky should have been preferentially located near the Galactic plane (as is seen with SGRs). If associated with older stars in the roughly spherical "bulge" of the Milky Way, GRBs would have been preferentially located in the direction toward the Galactic center and less so toward the opposite direction.[21] The inference that the Sun was roughly at the center of the GRB distribution in space, while casting aside some models, still allowed for a variety of distance scales: from a fraction of a light year to billions of light years.

The *brightness distribution* of GRBs appeared to show that we were seeing out to the edge of the GRB population: there were too few faint GRBs relative to the number expected if GRBs were uniformly ("homogeneously") distributed in space. Brightness was most straightforwardly measured as the peak flux (P, with units [erg s^{-1} cm^{-2}]) in the light curve of a GRB. The brightness distribution is usually measured as the number, $N(>P)$, of GRBs brighter than some peak flux P per year. If the peak luminosity (L, with units [erg s^{-1}]) of all GRBs is the same, then, using the $1/r^2$ law, for a given flux P we would see all the GRBs within a maximum distance:

$$d_{\max} \approx \sqrt{\frac{L}{4\pi P}} \propto P^{-1/2}. \qquad (1.1)$$

All the GRBs to that distance would be brighter than P by construction. The number of GRBs we would detect

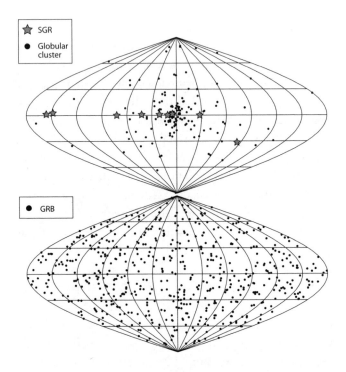

Figure 1.3. Distribution of different classes of objects and phenomena on the sky. Both top and bottom are projections of the celestial sphere such that the center of the projection is the direction toward the center of the Milky Way (in the constellation Sagittarius) and each line of constant latitude and longitude are spaced at 30-degree intervals. (top) Globular clusters, which appear diffuse, but centered around the Galactic center. The locations of the nine known or suspected SGRs (as of July 2010) are also shown (star symbol). All but the March 5th SGR (which is in the LMC), are within 6 degrees of the Galactic plane (represented by the longest horizontal line). This strongly indicates that the progenitors of SGRs are related to objects in the disk of our Galaxy. (bottom) In contrast, the distribution of GRBs (taken from the first few years of the BATSE experiment) appears randomly distributed throughout the celestial sphere; that is, GRBs appear to be isotropic. Adapted from B. Paczyński, PASP **107**, 1167 (1995).

to that brightness (or brighter) in one year would just be the volume times the intrinsic rate ($\mathcal{R}$, in units of [event yr^{-1} per volume element]): $N(>P) \propto V \times \mathcal{R} \propto \mathcal{R} \times d_{max}^3 \propto \mathcal{R} \times P^{-3/2}$. So with a homogeneous distribution, we expect that the number of faint GRBs should grow as a powerlaw proportional to $P^{-3/2}$, where the constant of proportionality scales directly with the intrinsic rate $\mathcal{R}$: for every ten times fainter in flux we observe, we would nominally expect about thirty-two times more GRBs.* While this was indeed seen for the brightest events, there was a flattening at the faint end of the brightness distribution. This flattening was highly suggestive that we were seeing the "edge" of the GRB distribution in space, an important clue in understanding the distance scale. But without knowing the intrinsic luminosity L, we could only infer the shape of the distribution, not the scale. It was like seeing a picture of a building but not knowing if it was of a miniature in a snow globe or the life-sized version.

One of the primary criticisms of the interpretations of apparent isotropy and inhomogeneity was that the population of GRBs observed thus far were observed with a variety of instruments on different satellites. Each detector had its own range of sensitivity to GRBs in energy and duration. One GRB might trigger one detector but not get recorded by another. Even in the smaller subsets of GRBs detected by a single instrument, where some of the concerns of a heterogeneous dataset could be mitigated, fainter events were not detected with as high efficiency as

---

*Plotting the logarithm of $N(>P)$ versus the logarithm of P would show a distribution following a line with slope $-3/2$.

brighter events; detector threshold effects were a natural explanation for the apparent paucity of faint events. The robustness of the isotropy measurements were also in question: what if the instrument spent more time looking in one direction than another? In that case, the observed distribution on the sky would not represent the true distribution. By 1990, given these observational caveats, there was enough wiggle room in the global statistics to allow for almost any distance scale and hence a broad range of progenitors.

To settle the questions about the peculiarities of the data collected over multiple telescopes, a single experiment was required, whose sky coverage and trigger efficiencies were well modeled, capable of detecting GRBs to significantly fainter levels than before. Such an experiment had been in construction at Marshall Space Flight Center (in Huntsville, Alabama) throughout most of the 1980s, having been accepted for funding by NASA in the late 1970s. The *Burst and Transient Source Experiment* (BATSE) was launched by space shuttle *Atlantis* on April 5, 1991, on board the *Compton Gamma-Ray Observatory* (CGRO; along with three other high-energy experiments). BATSE was about ten times more sensitive than previous GRB missions, allowing it to find roughly one burst a day for what would be its nine-year mission. BATSE also had a unique configuration on the spacecraft that allowed it to localize individually the 2,700 GRBs it discovered: the BATSE detectors were placed at the eight vertices of CGRO, allowing at least three detectors to "see" any place in the sky at one time. Since the intensity of light is proportional to the angle it makes with a detector,[22] the relative GRB flux recorded in the detectors could be used to triangulate the location of the GRB on the sky. For the first time, a single satellite would be capable of positioning a GRB to ~10 degrees on the sky. Moreover, the efficiency for discovery of GRBs in different places on the sky was well understood,[23] allowing for the cleanest reconstruction of the true spatial distribution of GRBs on the celestial sphere.

Given the strict criteria for triggering on a GRB, the BATSE efficiencies for detecting faint bursts were also well understood. The result was a remarkable and clean vindication of the two emergent properties of the GRB population. In the first few years of operation, the statistical consistency of isotropy of GRBs on the sky was confirmed beyond all reasonable doubt, as was the rollover of the brightness distribution. Any viable model for GRBs would need to place the Sun at (or very near) the center of a GRB population in space where the apparent rate of occurrence of the farthest events was less than those nearby.

Only two progenitor distributions survived the brutal BATSE scalpel. The first was a population of progenitors in the outer halo of the Milky Way, producing GRBs at such appreciable distance (~100 kpc) that the ~8 kpc offset of the Sun from the Galactic center would not be noticed by BATSE as a slight anisotropy. The paucity of faint events would be explained by the finite volume of the Galactic halo. The other viable scenario was a distribution of progenitors associated with galaxies billions of light years from the Milky Way. The isotropy would then be a natural consequence of the nearly perfect homogeneity of the Universe on large scales in all directions. This homogeneity had been observationally confirmed with the measurement of isotropy of radio *quasars*, bright and distant massive *black holes* (BHs) at the centers of distant galaxies. The brightness distribution of GRBs could be explained both by the finite age of the Universe and by effects due to the expansion of the Universe itself.

The results from BATSE nucleated two schools of thought: either GRBs were from "extragalactic" distances, and hence immensely and almost unfathomably bright, or from less extreme "Galactic" distances. The general consensus before the BATSE results was that GRBs were of a Galactic origin, but the BATSE results swayed many to the extragalactic camp. Each school had its vocal advocates. To commemorate the seventy-fifth anniversary of the famous 1920 Herber Curtis and Harlow Shapley debate over the then-controversial size scale of the Universe, a second "Great Debate" was held in 1995 in the very same auditorium as the first debate in the Natural History Museum in Washington, D.C. Arguing for the Galactic distance was Donald Q. Lamb from the University of Chicago. Arguing for the extragalactic distance scale was Bohdan Paczyński (pronounced "Pah-chin-ski") of Princeton University. After more than an hour of point and counterpoint, a poll of the GRB-aficionado-laden audience in the crowded room showed a rough split between the scenarios.[24] The crux of Paczyński's argument was that, regardless of the details, the most natural explanation for the isotropy and inhomogeneity was an extragalactic origin. Lamb was more tactical, noting for each relevant observation how a Galactic origin *could* accommodate the data. Most important for the Galactic argument was that, thanks to the March 5th event, we had an existence proof: we knew that at least some types of neutron stars were capable of making at least some types of gamma-ray transients. Dating back from the 1970s, there were some ideas of what could produce GRBs at extragalactic distances, but none was a particularly mature theory, and none was well tested by observations of the day.

In the social event immediately after the Great Debate, I, a precocious undergraduate student and fond of the Galactic picture, challenged Prof. Paczyński on some of the finer parts of his argument (which, of course, he parried well). He ended our conversation with a story about how most data had for years been showing that the Milky Way had a bar-like structure in the center, like many other galaxies, but that it took the death of the main proponent of the "no-bar" hypothesis for global consensus to glom on to the bar hypothesis. That albeit well-respected proponent had failed to convince the next generation of scientists why his story was compelling in the face of the data. As Professor Paczyński admonished, controversy and uncertainty in science are never settled by just talk, let alone a single debate. They are settled by new observations and insights about those observations. Eventually the "incorrect" view of the way in which the Universe operates is simply shown to be wrong, and all the remaining skeptics either change their minds or die.

1.6 The Afterglow Era Begins

It was abundantly clear in 1995 that new observations would be needed to settle the distance scale debate once and for all. Just as always, the main hope remained to find a precise location of classical GRBs and associate those positions with known objects. Since the single-satellite precision of BATSE was about as good as that detector configuration could provide, the key was to find counterpart events at other wavebands. It was known from precise IPN locations that no long-lived (weeks to months) optical transient accompanied the events. But thanks to the nearly instantaneous relay of new GRB positions from BATSE, several groups began construction of optical telescopes awaiting email alerts from GRB satellites; this new generation of telescopes was capable of rapidly observing new GRB positions. If GRBs were accompanied by optical flashes lasting just a few minutes, the robotic "on-call" telescopes would have a chance of catching the fleeting light. Theory of GRB emission mechanisms, developed in the early 1990s, also predicted radio transients from GRBs (chapter 3); as such, some groups began a search at radio wavebands with premier facilities like the *Very Large Array* (VLA) in New Mexico. None of the early searches of BATSE positions proved fruitful.

A group from the University of California, Massachusetts Institute of Technology (MIT), and Los Alamos proposed a new satellite mission in the mid-1980s that would seek to localize GRBs using the X-rays that where known to accompany many events. In what shaped up to be a multinational collaboration with France and Japan, the *High-Energy Transient Explorer* (HETE) was designed to trigger on new GRBs using an instrument without positional capability, then use another instrument to localize the X-rays from that GRB. The concept behind the X-ray camera was to create a mask pattern that would block some of the incoming X-ray photons but let the others fall on a grid of position-sensitive X-ray detectors. The mask would create a distinct shadow depending on the direction of the GRB, which in turn could be used to reconstruct the location of the event to IPN-level accuracies (tens of square arcminutes uncertainty regions) but within a few seconds of the trigger. Unfortunately, HETE was lost shortly after launch in 1996, but the second incarnation, HETE-2, was put together with mostly spare parts and launched in 2000; HETE-2 would quickly vindicate the new X-ray localization approach.

While GRBs were not the main scientific priority of a new Italian-Dutch experiment, the "Satellite per Astronomia a Raggi X" (commonly known as *BeppoSAX*), it did carry a Gamma-Ray Burst Monitor (GRBM) that stared at the same place on the sky as the X-ray-coded mask imagers (collectively called the Wide-Field X-ray Camera [WFC]); those instruments were capable of seeing roughly 1/15th of the sky at any time. Launched in April 1996, BeppoSAX demonstrated that summer that a burst triggered in the GRBM and seen in the WFC field of view could be localized to ~5–10 arcminute radius. Such localization accuracy was often obtained with the IPN (see §1.3), but BeppoSAX had the ability to determine such positions in a matter of a few hours, whereas IPN localizations typically took days to determine. The speed of finding a precise GRB position would prove to be a crucial capability.

On February 28, 1997, BeppoSAX localized a GRB using the WFC and the GRBM and found a good-enough position to command the satellite to repoint to allow its

Narrow-field X-ray Instruments (NFI) to image the GRB region. Just eight hours after the GRB, two of the NFI cameras detected a bright, new X-ray source consistent with the WFC position. Three days later, that source had vanished.

Not only had the BeppoSAX team found the first *afterglow* of a GRB, the X-ray afterglow was positioned well enough (to 50 *arcseconds* in radius) to allow for sensitive searches of counterparts at other wavelengths. Just twenty hours after the GRB, a group led by Jan van Paradijs at Amsterdam University, the Netherlands, obtained a set of "deep"* optical images of the WFC error location from La Palma Observatory on the Canary Islands (Spain). One week later, my collaborator (Nial Tanvir) obtained the next set of images of that field from the same observatory.[25] A faint source in the NFI error box had vanished between those two epochs: it was the discovery of the first fading optical (i.e., visible-light) afterglow of a GRB. The optical position, determined to better than 1 arcsecond, was by far the best localization of any GRB, including the March 5th (SGR) event.

As the optical afterglow faded, the *Hubble Space Telescope* (HST) was trained on the position and found a blue blob around the fading afterglow. To most, this "nebulosity" looked a lot like a faint low-surface brightness galaxy, thus confirming the extragalactic hypothesis; but to the stalwarts of the Galactic model, it looked like an NS-blown

---

*The adjective *deep* is often used to described long exposures on the sky, where fainter and fainter objects can be detected with higher and higher signal-to-noise; these fainter images probe a larger *depth* of the Universe than "shallower" exposures.

bubble (confirming the opposite). Despite some claims from a few that the afterglow appeared to be moving on the sky (a nominal expectation of the Galactic model) and that the nebulosity changed color and shape (only possible in the Galactic model), all doubts about the distance scale of classical GRBs would be soon laid to rest.

On May 8, 1997, BeppoSAX spotted another GRB, and an optical afterglow was promptly discovered consistent in location and time with another fading X-ray afterglow. Astronomers at the California Institute of Technology (Caltech) obtained a *spectrum* of the optical afterglow using the Keck II 10-meter telescope in Hawaii; this telescope (and its twin Keck I) was the largest optical telescope in the world at the time. A quick inspection of that data led to one of the biggest discoveries in modern astrophysics: notched out of an otherwise smooth *spectrum* were a series of *absorption lines*, characteristic of iron and magnesium in gaseous form. But instead of seeing lines at the specific wavelengths as they would appear in a laboratory, they all appeared shifted to significantly redder wavelengths.* This

---

*This effect is called *redshift*, and the amount of redshift is usually given with the symbol z. The relationship between the observed wavelength λ_o and the emitted wavelength of light λ_e is $\lambda_o = \lambda_e \times (1+z)$. Similarly, since $\lambda\nu = c$, the relationship between the observed frequency and the emitted frequency is $\nu_o = \nu_e \times (1+z)^{-1}$; so frequency decreases and wavelength increases with increasing value of z. If we wanted to associate the redshift of a source with its apparent velocity ν_{app} away from us, we can use the equation $1+z = \left(\frac{1 + v_{app}/c}{1 - v_{app}/c}\right)^{1/2}$, where c is the speed of light. Within the Milky Way, objects show redshifts (and even blueshifts, with $z < 0$) which are very small ($|z| \lesssim 10^{-3}$, where $|v_{app}| \lesssim 300$ km/s) but objects at large distances from the Milky Way can show large redshifts, $z > 1$. The next footnote discusses the various mechanisms leading to the apparent redshift of light.

redshift effect was immediately recognized as an effect due to the expansion of the Universe,* and, given a reasonably well-prescribed mapping between the redshift measurement and distance, this observation established that GRB 970508 must have occurred from a distance of more than about 5 billion *parsec* (i.e., 5 *Gpc*) away. Some gas cloud in a distant galaxy lay between us and the GRB,[26] and its absorbing metals provided enough of a unique fingerprint to measure an unambiguous redshift.

In one fell swoop of the telescope, the thirty-year marathon to measure the distance scale of classical GRBs was won. The cosmological distance scale, the disfavored choice of so many for so long, had triumphed.[†] To top it off, the first radio afterglow of a GRB was found following GRB 970508. A detailed inspection of the way in which the radio afterglow behaved, coupled with the then-known distance, strongly suggested that the afterglow-emitting surface was expanding at a rate close to the speed of light.

---

*An expanding universe tends to stretch out the observed wavelengths of light, causing redshift. The further away a source is on cosmological scales, the larger its apparent redshift. There are two other explanations for the origin of redshift. First, redshift can be due to relative velocity differences between the absorbing (or emitting) source and the observer. When moving apart, this redshifting is called a *Doppler shift*. Second, when light passes near any object with mass, its wavelength changes depending on the distance to the mass. Light emitted near the surface of a neutron star, for instance, is perceived to be more red by observers at progressively larger distances from the neutron star. This effect is called *gravitational redshift* and is a manifestation of General Relativity. In the case of GRB redshifts, no corroborating evidence for either of these two explanations are viable, leaving only the cosmological-expansion explanation.

[†]Recalling the discussion of the uncomfortable energetics that a cosmological distance scale would require, we might note Galileo's admonition: "Facts which at first seem improbable will, even on scant explanation, drop the cloak which has hidden them and stand forth in naked and simple beauty."

This *relativistic expansion*, which we will explore in §2.2.1, was a basic prediction of most cosmological theories of GRBs—another triumph for observers and theorists alike.

The seminal afterglow discoveries of GRB 970508 were first presented at a workshop less than two weeks later, conducted on the island of Elba, Italy. Franco Pacini and the BeppoSAX collaboration had organized that workshop following GRB 970228 and, as a student involved in the first optical afterglow discovery of GRB 970228, I was immensely thrilled to attend. Having not been part of the GRB 970508 results, I was honored to have been asked along with veteran theorist Mal Ruderman by the British journal *Nature* magazine to report on the Elba meeting in a *News & Views* article that accompanied the seminal papers on the GRB 970508 discoveries. We ended our article with a reflection on the accomplishment and a speculation on the future of the field: "It is hard to point to any other such important astrophysical problem that should be so quickly solved as the origin of gamma-ray bursts. But that will probably only be the first extraordinary chapter of a book that promises to become even more exciting."

1.7 Progenitors and Diversity

By the end of 1998, more than twenty GRBs had been rapidly localized to the several-arcminute level, and six had confirmed cosmological redshifts. So, while the distance scale was indeed confirmed, understanding the "origin of gamma-ray bursts," as we had perhaps zealously overstated in our article, was far from solved. With redshifts now

in hand, however, the energetics of individual events
could provide an important constraint. The total energy
release* of most events was in the range of 10^{51}–10^{52} erg,
comparable to the energy release in a supernova ($\sim 10^{51}$
erg) and reasonably consistent with a variety of progenitor
models where no more than a small fraction ($\lesssim 10^{-2} M_\odot$)
of stellar mass of energy† is released in the event.

However, the implied energy release of GRB 971214
threw the community for a major loop. Shri Kulkarni
and his collaborators (including me) at Caltech discovered
a high redshift of $z = 3.42$ (hence a very large distance)
for GRB 971214; the simplistic calculation suggested that
the gamma-ray energy released during the GRB—over the
course of just a few seconds—amounted to the equivalent
of an appreciable fraction (> 10 percent) of the entire
mass of the Sun! News outlets ranging from the PBS
News Hour[27] to the Drudge Report proclaimed the event
to be the "biggest bang in the universe recorded," with
an "intensity second only to the Big Bang." Hate email
from theorists, worried about observers overturning the
tidy stellar-mass progenitor hypotheses, piled up in Kulkarni's inbox. A relaxation of those uncomfortable energy
requirements (and the tempers of theorists) would come
the following year, as it was recognized that GRBs could

---

*Assuming that the burst emitted the same amount of energy in all directions.
See §3.3 for a discussion.

†For explosions that involve nuclear, rather than chemical reactions, it is natural
to describe the energy release (E) in mass-equivalent (M) quantities, using the
conversion of $E = Mc^2$ from Relativity Theory. Here c is the speed of light
and $M_\odot$ is the mass of the Sun. The Sun has an equivalent rest-mass energy of
1.78×10^{54} erg.

be collimated (or "jetted"). The implication was that the inference of energy release was too high. The geometry of GRB explosions is discussed more fully in §3.3.

On the opposite end of the energy spectrum, an otherwise normal-looking GRB (from a high-energy perspective) was discovered by BeppoSAX on April 25, 1998. Titus Galama, in the Netherlands, noted a curious bright source in the WFC error box, apparently in a spiral arm of a nearby[28] galaxy. Instead of the rapidly fading optical afterglow of other GRBs, this source got brighter with time. Later recognized as a peculiar and bright supernova (designated SN 1998bw*), the association both in place and time of this counterpart with GRB 980425 implied an energy release for the GRB several orders of magnitude smaller than the other GRBs previously studied. Nature, in producing the oddity that was GRB 980425, had provided another clear insight into a GRB progenitor. As the March 5th event showed us that NSs were capable of making some forms of GRBs, we now knew that massive stars (the likely origin of SN 1998bw) were also capable of making GRBs. Unfortunately, no convincing case of another extremely low-luminosity, 980425-like GRB has been found to date.† The question of whether GRB 980425 resides on

---

*Supernovae are named in the order they are officially discovered in a given year, starting alphabetically with the designation "A." After twenty-six SNe are found in a given year, two lower-case letters are used, starting with "aa." SN 1998bw was the seventy-fifth SN found in 1998.

†To be sure, there have been a number of "underluminous" GRBs that bridge the energy release between GRB 980425 and other cosmological GRBs. GRB 060218 and GRB 100316D, both at low redshift and associated with supernovae, had total gamma-ray energies about 10–100 times more than GRB 980425. See, for example, R. L. C. Starling et al., *arXiv/1004.2919* (2010).

the extreme tail of the "classical GRB" set or represents a physically distinct class of GRB-producing progenitors is still not a settled question.

Observations of the galaxy hosts of GRBs and the locations of GRBs around those galaxies provided strong circumstantial evidence that classical GRBs were more associated with younger stars than older stars. Primarily on two grounds this was troublesome for the popular view was that cosmological GRBs arise from the merger of NS binaries. First, many scenarios for NS coalescence posited a long delay (approximately one billion years) for the event after the birth of the progenitor stars. In that time, the star formation in that galaxy could subside. So why GRBs were all associated with blue galaxies* (the observational hallmark of active star formation) was a mystery. Second, NS binaries were expected to travel far from their birthsite before coalescence, so GRBs from such progenitors would have no business being associated spatially with star formation.

The first direct evidence for a massive-star progenitor of a cosmological[29] GRB came with my discovery of a curious supernova-like feature in the late-time light curve following GRB 980326.[30] Many other "SN bumps," some exhibiting supernova features to a high degree of confidence, were subsequently detected in other GRBs; this led to the speculation that most, if not all, classical GRBs could be connected to massive star explosions, rather than NS binary mergers. However, it was the

---

*This was true from 1997 to 2005. Even today only a minority fraction of GRBs are associated with older galaxies. See chapter 4.

spectroscopic observations following GRB 030329, first
by groups at Harvard University and in Copenhagen, that
the "smoking-gun" evidence directly connecting classical
GRBs to massive stellar death was uncovered. Embedded in the spectrum of the afterglow light of that GRB
were telltale supernova spectral features. Amazingly, that
supernova strongly resembled SN 1998bw, but unlike
GRB 980425, this event occurred at a "cosmological"
distance.

Both SN 1998bw and SN 2003dh (as the SN associated
with GRB 030329 was designated) were of an unusual
type of broad-line "Type Ic" SN, thought to arise from
a star once more than 30–40 times the mass of the Sun but
having expelled its hydrogen and helium envelopes as it
neared the end of its life cycle. These are some of the most
extreme SNe in nature coming from some of the most
massive (and rare) stars. Massive stars are like the James
Dean of the stellar population: they live fast and die young.
Though they have much more fuel to burn than stars of
mass comparable to the Sun, massive stars burn fuel at
much higher rates than low-mass stars. So, when a group
of stars of different masses is formed, the most massive stars
expend their fuel first, typically within 10–50 million years
after nuclear burning begins. Hence, when we say that
GRBs are connected with young stars and ongoing star
formation, we are also implying a connection to massive
stars.

Broadly speaking, the association was a wonderful full
circle for the field. The first theory of GRBs (before
GRBs had been discovered) involved supernovae, and
it was the association with SNe that Klebesadel and

his collaborators first tested.* However, the mechanism whereby the supernova *shockwave* itself would produce a GRB cannot account for the energy release in most GRBs. Instead the observations were more consistent with a progenitor scenario—commonly referred to as the "collapsar" model—first put forward by Stan Woosley in 1993.[†] In the current collapsar picture, a fast-moving jet is launched from a newly created BH at the center of a collapsing massive star. The GRB is produced as material in the jet shocks against itself. The SN is created as heat released from freshly minted radioactive elements powers[‡] the explosion of the star.

Supernovae have also been manifest after several other GRBs since 2003, but, like the GRBs themselves, there appears to be a large range in the basic properties of the SNe (energy releases, velocities, synthesized radioactive mass, etc.). In a challenge to the notion that all classical GRBs are from massive stars,[§] two low-redshift GRBs were found in 2006 that did not have accompanying SNe to

---

*We know now that their null result was due to a mismatch of the (large) distances of GRBs and the more nearby supernovae found in the 1960s and 1970s.

[†]This original scenario failed to produce a supernova concurrently with the GRB but it was later refined.

[‡]Radioactive material is, by definition, unstable. Over time the nuclei of radioactive material lose protons and neutrons as that material transforms into different flavors of the same element (called "isotopes") and to other elements altogether. In this process of decay, energy is released; and, since different isotopes decay at different rates, the power available due to radioactivity depends sensitively on the original composition of the radioactive material.

[§]To be sure, most GRBs do not appear to be accompanied by supernovae-like events. However, most detected GRBs originate from $z > 1$ where no SN features could have been discernible by current follow-up facilities: essentially, any SN that far away would be too faint to detect.

very deep limits (see §5.1.3). Still unsettled is whether these SNe-less GRBs represent yet another progenitor mechanism or simply are a manifestation of the original collapsar scenario where too little radioactive material is created to power an observable supernova.

The discoveries and insights that accumulated for the first eight years of the afterglow era were relevant to the majority of classical GRBs. Known since the 1970s was that a minority of events were especially short in duration and yet not apparently SGR flares. In 1992, observations with BATSE showed two distinct types of GRBs: the short events, those lasting less than about two seconds, appeared on average to have a "harder" spectrum than the long events (see figure 2.4 in the next chapter). By 2004, a bona fide "short-hard" GRB (SHB) had never been localized well by BeppoSAX or HETE-2 in part because the gamma-ray detectors were significantly less sensitive than BATSE to short-duration events. All the same questions persisted. Were the afterglows of SHBs the same as from the majority "long-soft" bursts (LSBs)? Was the distance scale also cosmological? Were SHBs also due to massive stellar death?

On May 9, 2005, the new NASA GRB satellite called Swift,[31] responded autonomously to a short burst that had triggered its onboard gamma-ray detector called the *Burst Alert Telescope* (BAT). Swift was a novel facility in that BAT-discovered GRBs would not only be recognized quickly (<few seconds) by onboard computers and the discovery broadcast to the ground but also that the satellite itself would respond to the event by autonomously repositioning itself to capture the new GRB position with X-ray and ultraviolet/optical imagers. Pointing its

X-ray Telescope (XRT) at the GRB localization, Swift astronomers discovered the first X-ray afterglow of a short burst that day. With the precise X-ray location promptly relayed to the world, telescopes big and small were trained on the (<10 arcsecond) X-ray position. While no optical afterglow was found, the X-ray position was curiously close to a massive red galaxy containing little-to-no ongoing star formation. Though the physical association of GRB 050509b was not unassailable, the spatial coincidence with a "red-and-dead" galaxy was a remarkable departure from what was seen for long-burst galaxy associations. My group based at the University of California, Berkeley (and others) argued that the GRB must have originated from an older population of progenitors than most LSBs; and, given the appreciable physical offset of the X-ray position from the galaxy, one of the basic predictions of the NS-merger model had been borne out by those observations.[32]

More than two dozen SHBs have been well localized since the summer of 2005, and the verdict is still out on their origin and even the rate of SHBs throughout cosmic time. As will be discussed in chapter 5, while not all SHBs are associated with red galaxies, the inference that they are due to an entirely *different* class of progenitors than that of LSBs seems to hold up. Unlike the collapsar scenario, which posits that a certain class of supernova should occur contemporaneously with a GRB, if the merger model is correct for SHBs, then a definitive observation confirming the model will be very difficult to obtain. The only obvious smoking gun for such progenitors—because they involve the violent coalescence of compact masses—would be the detection of *gravitational waves* coincident with the GRB (see §5.2.3 and §6.4.1). Detection of such an event is a primary astrophysical impetus for the construction of next-generation gravitational-wave observatories capable of detecting NS mergers to large-enough distances in space.

1.8 Gamma-Ray Bursts in a Universal Context

Throughout this book, we explore our understanding of the diversity of GRB progenitors and the physics that gives rise to the prompt high-energy emission and the afterglows. Some of this understanding is a work in progress. Figure 1.4 shows some extragalactic progenitors that appear not only plausible but very likely to be involved in producing GRBs. Observational confrontation with theory has already opened up new vistas on the life cycle of massive stars and the nature of *radiation* processes from fast-moving material.

Yet irrespective of the theoretical underpinnings of the events themselves, there are some robust inferences about GRBs that hold vast implications for astrophysics as a whole. First, GRBs are for a brief time the brightest events in the Universe across the electromagnetic spectrum. At optical wavelengths, some of the brightest GRB afterglows are more than ten thousand times brighter than the brightest known quasar in the Universe and millions of times brighter than any supernova ever observed. Second, they occur in and around galaxies throughout the history of the Universe. Last, the progenitors that make GRBs were producing GRBs less than one billion years after the Big Bang. All these allow GRBs to serve as unique probes

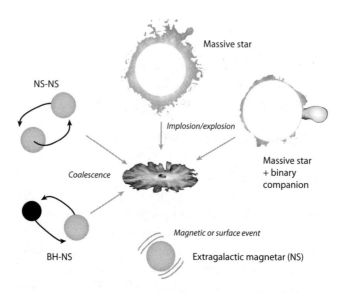

NS-NS

Massive star

Implosion/explosion

Coalescence

Massive star + binary companion

BH-NS

Magnetic or surface event

Extragalactic magnetar (NS)

Figure 1.4. Schematic of the various viable progenitors of cosmological GRBs. All involve the sudden release of energy in a small volume of space, triggered by the mechanisms noted in italics. Aside from extragalactic magnetars, the GRB arises from the rapid conversion of mass to energy. A black hole (BH) at the center of an *accretion disk* of in-spiraling material (shown at center) lies at the heart of many scenarios for energy liberation. The "central engine" of GRBs will be discussed in more detail in §2.3. Progenitor scenarios are discussed in chapter 5. The progenitors sizes are not shown to scale.

of other grand questions about the Universe that are not directly related to the events themselves.

While the afterglow light is bright, GRBs make for exceptionally useful lampposts: the light penetrates the gas and dust intervening in the line of sight from the burst location to the detector. Recall that the redshift of

GRB 970508 was found using absorption lines by gaseous metals. The detailed study of those lines, sometimes due to gas in galaxies that are spatially distinct from the GRB region, has started providing interesting views of the chemical state of galaxies in the distant past and information on how the chemical enrichment of galaxies is changing with cosmic time. The most distant GRBs hold the promise of revealing the state of the Universe itself, allowing us to measure the way in which the gas in the Universe transitioned from being in a fully neutral state to a fully ionized state. By the end of the first decade of 2000, GRBs were the redshift record holders, having catapulted past the most distant known quasars and galaxies with the discovery of a $z = 8.2$ event (GRB 090423). We explore in chapter 6 the ways in which GRBs are reinvigorating other pursuits in astronomy as well as opening up new possibilities to test long-held beliefs, such as the notion that the speed of light is constant in vacuum.

Like any maturing scientific endeavor, the development of an understanding of the phenomena that initially draws in the lessons from a variety of different disciplines eventually leads to new tools to study the world in a larger context. Along the way, there are triumphs and missteps. But, like all science, the story is never complete—we can only hope to close the chapters in front of us as we open up new ones.

2

INTO THE BELLY OF THE BEAST

There are so many more questions yet to be answered... And so I wonder,... Are we alone in the universe? **What causes gamma-ray bursts?** What makes up the missing mass of the universe? What's in those black holes, anyway? And maybe the biggest question of all: How in the wide world can you add $3 billion in market capitalization simply by adding .com to the end of a name?

—William Jefferson Clinton,
Science and Technology Policy Speech,
California Institute of Technology, 21 January 2000
(forty-nine days before the peak of the NASDAQ stock exchange and the start of a ten-year bear market)

2.1 What Are Gamma-Ray Bursts?

Before the afterglow era, GRBs were essentially defined by observations of their high-energy emission.* The landscape of such observations—the light curves and spectra of the events—exhibits at once great diversity and elements of commonality that bind different events together. As we shall see, GRBs are like fingerprints: no two are alike, but they share common properties. Those common elements provide strong constraints both on the nature of the

---

*For the purposes of this discussion, we will take high-energy emission to be any light observed with photon energies larger than 10,000 *electronvolts* (*eV*)—that is, energy in the X-ray and gamma-ray portions of the electromagnetic spectrum.

"engine" that supplies the energy to the event and the physical processes that drive the emission we see.

2.1.1 Light Curves and Spectra

The community has been fortunate to have had continuous GRB monitors in space since the discovery of the phenomenon. BATSE became the workhorse for much of the 1990s, triggering on 2704 GRBs (about one GRB per day for nine years). Combined, BeppoSAX, HETE-2 and Integral[1] observed more than one hundred events in the late 1990s and early turn of the century. Since the end of 2004, Swift has been discovering GRBs at a rate of about two per week. A number of other satellites (Fermi, Ulysses, Konus-Wind, Suzaku) also contribute to the overall discovery rate. By 2010, the number of observed GRBs approached five thousand.

Figure 2.1 shows a sampling of GRB light curves from BATSE. The roughly ten-second duration of the events above the background level coupled with subsecond variability is only a broadly acceptable characterization of the population. In detail, some events are shorter in duration, and many are much longer. Some events appear to have a single pulse of emission, falling more slowly than during the rapid rise, and some have multiple pulses. Some pulses appear to rise more slowly than they fall. The widths of pulses vary between bursts and within bursts. Some have an initial complex of pulses, then a long period of quiescence before another complex. Some show faint "precursor" events before the bulk of the emission is seen.

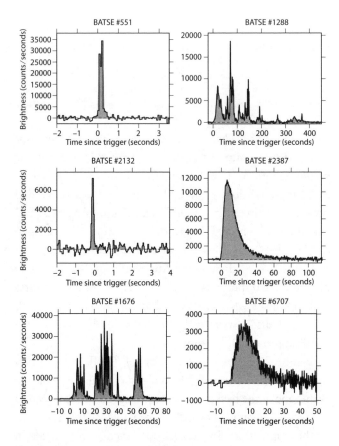

Figure 2.1. Panoply of GRB light curves, as observed by BATSE. Events #551 and #2132 are short-duration events, and all others shown are long-duration events. Some events are smooth, apparently a single pulse (e.g., #2387 and #6707), and others are more spikey (#1676 and #1288). The event #6707 is the gamma-ray light curve of GRB 980425 (see §1.7).

Aside from SGR flares, no classical GRB appears to have strong evidence for periodicity in the light curve.

Making sense of this light-curve zoo is not a trivial exercise. To date there is no consistent physical model for GRB light curves that can explain all the properties of all GRBs (see §2.3.3). Nevertheless some of the global properties of the events are routinely measured and compared with each other. Most prevalent in the literature are measures of duration, both of individual pulses and of the totality of the activity. For an isolated pulse, the width may be taken as the time in seconds between when the pulse reaches 50 percent of its peak brightness and then falls back to 50 percent of that brightness. With only a finite number of photons collected and in the presence of noise, it is impossible to measure this value precisely, so the measurement of pulse widths has some inherent degree of uncertainty. The duration of an event is typically taken as the total time between when 5 percent and 95 percent of the total energy* above the background level is accumulated. This duration, in units of seconds, is referred to as T_{90} because it counts the range over which 90 percent of the fluence is detected.

The spectra of GRBs, showing us how the photon energies are distributed, also exhibit differences between events but, unlike light curves, the vast majority of measured spectra appear to be well fit by a simple empirical model. As one might expect, most of the energy release in GRBs occurs in the gamma-ray regime, above about 10 keV.

---

*This is referred to as *fluence* and has units of energy per collecting area. Mathematically, fluence is the integral of the light curve flux over time.

Figure 2.2 shows the spectrum of a bright GRB observed with multiple instruments on the Compton Gamma-Ray Observatory. The dashed line shows a four-parameter empirical fit called the Band function.[2] The parameters are the peak energy (E_{peak}), the brightness at the peak, and the two powerlaw indices at energies below and above the peak (α and β, respectively). The values of E_{peak} typically range from 10 keV to a few MeV. The values of α are clustered around -1, and the typical value of β is -3 (it is straightforward to show that, for the energy to remain finite, β must be less than -2). Not surprisingly, the *bandpass* of the detector severely biases the clustering of the observed E_{peak} values—your eyes only let you see what they are sensitive to; indeed, the *Ginga* experiment, with a lower-energy bandpass, saw many more soft (i.e., low E_{peak}) GRBs than BATSE (figure 2.3). For GRBs that are bright enough to take a snapshot spectrum throughout the event, all four parameters are seen to change with time. Though there are plenty of exceptions, in general E_{peak} moves from larger to smaller values during a GRB, exhibiting so-called "hard-to-soft" evolution.

To help with the discussions that follow, there are some general light-curve and spectral trends worth noting:

- **Pulse Width Evolution**: In bursts with multiple pulses, the fainter pulses tend to be wider in duration.[3] However, for pulses of about the same peak brightness, the widths are generally seen to remain constant throughout the GRB. That is, a pulse of a given peak brightness is likely to have

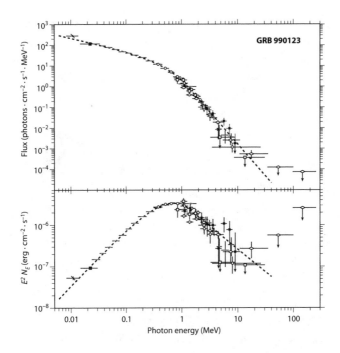

Figure 2.2. The time-averaged spectrum of GRB 990123. (top) The spectrum in units of photon flux (N_E) or brightness. This shows that most of the photons in a GRB are emitted at the low end of the energy range (into the hard X-ray bands). But the same spectrum in units of energy flux, as seen in the bottom plot, shows that most of the energy of a burst is emitted in the gamma-ray range. Here, this integrated spectrum of the first thirty-two seconds of the event is well fit by a Band function (dashed line) with a peak energy $E_{\text{peak}} = 720 \pm 10\,\text{keV}$, $\alpha = -0.60 \pm 0.01$ and $\beta = -3.11 \pm 0.07$. Various instruments and experiments on CGRO were used to construct the spectrum. Adapted from M. S. Briggs et al., *ApJ* **524**, 82 (1999).

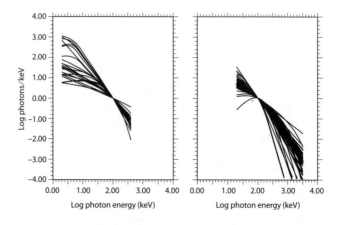

Figure 2.3. A compilation of GRB spectra as observed by *Ginga* (left) and BATSE (right). The Band-function fit spectra have all been normalized to one photon per keV at an energy of 100 keV. *Ginga* had a softer bandpass than BATSE and therefore detected systematically softer events. Both axes are shown as the logarithm of the quantity. Adapted from T. E. Strohmayer, E. E. Fenimore, E. E. Murakami, and A. Yoshida, *ApJ* **500**, 873 (1998).

the same duration regardless of whether it was the first pulse in the event or the last.

- **Pulse Widths in Energy**: Though the widths of pulses in a given GRB can be varied throughout the event, it is generally the case that individual pulses appear wider in duration at lower energies. The obvious implication is that the same GRB observed with different detectors will appear more or less spiky depending on the energy range where those detectors are sensitive. Not so obvious is that the same GRB placed at farther distances will appear not only fainter overall but more spiky.

The cosmological redshifting (see footnote *, p. 28) of the GRB brings into view a higher-energy portion of the GRB spectrum than would be viewed by us without the cosmological effect.[4] A counteracting effect is *cosmological time dilation*, which stretches out pulse widths.

- **Pulse Lags**: The time when a pulse peaks at lower energies relative to the peak time at higher energy is called the "lag" and is measured, usually, in tens of milliseconds. Lags are measured by noting the shift in time required to align a pulse peak in two different energy ranges. Most lags are somewhat positive (a manifestation of hard-to-soft evolution), many lags are consistent with zero (that is, a pulse peaks at nearly the same time in multiple bandpasses), and a few appear to have negative lag—where the softer peak precedes the harder peak. Lags never appear to be longer than about 10 percent of the T_{90} duration of an event.

- **Polarization**: There have been some claims that photons in the prompt GRB emission are "polarized," meaning that the electric field vectors of different photons are, in the aggregate, ordered rather than randomly distributed. If true, then emission from a source with coherently aligned magnetic fields would be the most natural explanation. The polarization measurements are very difficult, and the few positive claims are controversial.[5]

Complicating the derivation of even the most basic metrics, such as event duration, is that the specifics of what we observe depend greatly on the peculiarities of both the instruments themselves and the circumstances under which the measurements are made. For instance, a GRB will look different if observed when the instrument is experiencing different levels of background radiation.[6] There is a strong bias toward seeing only the brightest moments during the event: the higher the background, the less we see of the faintest emission from the burst.

2.1.2 Classification at High Energies

As seen in figure 2.4, there is clearly a distribution in the peak energy of GRB spectra. Those events with the largest ratio of energy observed in the X-ray band to the gamma-ray band—generally when E_{peak} is less than about 15 keV—are called "X-ray Flashes" (XRFs). Those with a comparable amount of energy in the two bands are referred to as "X-ray Rich" (XRR) GRBs.[7] Everything else is just referred to as simply a "GRB" (or for those who enjoy retronyms, "classical GRBs"). There is no clear delineation between these three spectral classes.[8]

The durations and aggregate spectral properties of GRBs do appear to cluster. We have known since the early 1980s that the duration distribution of GRBs (that is, the number of observed events with a certain T_{90}) is bimodal: the majority of GRBs last more than a few seconds, and a minority last less than one second.[9] A relative dearth of events with T_{90} of about one to a few seconds leads to the appearance of this bimodality. In a seminal paper based on the first two years of BATSE data, Chryssa Kouveliotou

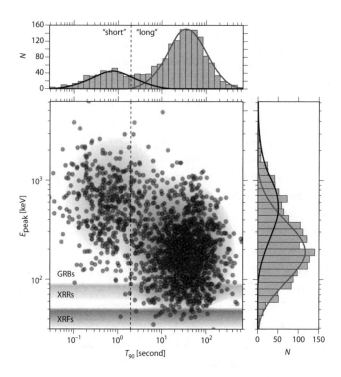

Figure 2.4. The spectral-duration distribution of GRBs as seen by BATSE. The scatter plot of E_{peak} versus T_{90} shows two clear loci of events (shaded regions) with shorter events appearing preferentially harder (i.e., larger E_{peak}). The approximate E_{peak} distribution of XRFs, XRRs, and GRBs is shown as horizontal gray regions. The outset histograms, showing the number of events in a certain parameter range (top: duration; right: peak energy) also show a separation (most prominently in duration). The smooth curves atop the histograms show representative log-Gaussian fits to the distributions. The traditional two-second dividing line between "short" and "long" GRBs is shown. While the bimodal distribution in duration persists across instruments, a plot of hardness

(*Continued*)

and her collaborators discovered that the longer-duration events also appeared to be softer on average.[10] Figure 2.4 shows the relationship between duration and hardness in which this trend is manifest. There have been some claims of a third and even fourth cluster in this space, but the statistical significance of such additional classes does not appear to be very strong.

Finding trends and relationships between observables is not only an important trait for human survival,[11] but it is also an important platform in making sense of an otherwise large melee of metrics on phenomena we care about. Indeed, taxonomy in science is not just stamp collecting but a bona fide tool useful in uncovering the physical relationship between members. The duration-hardness distribution (figure 2.4) serves as the backbone for GRB taxonomy. These two parameters are among the most easily observed in an event, and they clearly have the power to distinguish two broad classes of events.

As discussed in chapter 1, long-duration soft-spectrum GRBs (LSBs) have been associated with the death of young massive stars, while short-duration hard-spectrum GRBs (SHBs) have now been associated with an older stellar

---

Figure 2.4. (*Continued*) versus duration for Swift-only GRBs would appear differently: since Swift is relatively insensitive to detecting very hard events of short duration and very soft events at long duration, there is no clear bimodality in hardness of Swift events. The uncertainties in the measured quantities for each event are not shown for clarity. I am grateful to Nathaniel Butler who kindly provided the E_{peak} and T_{90} fits to the BATSE data.

population. So, at first blush, it would seem that this high-energy classification may indeed reflect a significantly different physical origin of the events. There are complications, however. First, even after accounting for various observational biases, ascribing membership of a given event to one class is inherently probabilistic since there is clearly a broad overlap between phenomenological classes (figure 2.4). A burst lasting one second certainly could belong to the tail end of the "long" class, while a burst lasting ten seconds could properly belong to the "short" class. Second, the same burst observed at different redshifts will have a different observed spectra and T_{90}. In particular, an SHB occurring at high cosmological redshift could easily manifest itself as an LSB. Likewise, some long-duration bursts with an overall soft spectrum could have a short/hard initial pulse. If faint enough, the long/soft portion of the burst could be lost in the detector noise, making the source seem like an SHB. Last, there appear to be many individual events from within the Milky Way and other nearby galaxies that could easily have been classified as an LSB or SHB based on high-energy properties alone but that arise from very different progenitors* than the majority of the events.

Whether the high-energy properties of GRBs can directly reflect the progenitor diversity is a question we will revisit in §5.4. The progenitor question as it relates to observables is, in some sense, like asking about the properties of performers singing in a chorus. Are all baritone parts sung by men? Is someone's degree of vibrato correlated

---

*For example, magnetars. See §5.3 for an extended discussion.

with his/her age or with the part of the world he/she was born in? Irrespective of such inquiries, we can ask what is the physical origin of the sound and voice itself, or, in our case, what is the physical origin of the light of GRBs.

2.2 Understanding the Origin of the High-Energy Emission

What is seen in the first moments of a GRB is commonly referred to as the "prompt emission"; as shown previously, most of the prompt energy is in the form of gamma rays. During the GRB itself, roughly 0.1 percent of the *restmass energy* of the Sun is thought to be radiated away in gamma rays alone. The spectra and light curves of GRBs provide vital clues to frame our understanding of the events. For instance, since gamma rays are a natural consequence of radioactive fission decay, it is reasonable to ask whether GRBs are due to such decay. Setting aside the challenges in arranging an event where so much energy is released from radioactive decay, this scenario is ruled out since we do not see evidence for fission-process gamma-ray lines in GRB spectra. Moreover, the light curves are not consistent with what is expected from radioactive decay.[12] What if, like in supernovae, the energy of the radioactive material was captured by the material in the explosion itself? In technical terms, we would say that the material had high *optical depth** to the absorption of gamma-ray light. In this

---

*Optical depth, usually symbolized as τ, is a dimensionless number describing the stopping power of matter to light at a certain wavelength of light. Physically,

scenario, we could expect to wash out the gamma-ray lines, and the energy otherwise carried in those lines would heat up[13] the absorbing material; this energy is then reradiated, usually at lower frequencies. The emergent spectrum of this reradiated energy would be expected to be that of a *blackbody*.* Unfortunately for the vast majority (if not all) of the observed GRBs,[14] it is not: bursts fit by a Band spectral shape are too bright at high energies and too faint at low energies to be consistent with a blackbody.

2.2.1 Compactness Problem and Relativistic Outflow

That GRB spectra are not consistent with blackbody emission implies that, at gamma-ray energies, the emitting system is optically thin (i.e., $\tau < 1$). This turns out to be a major constraint on the nature of the material that participates in the radiation. The most important part of

---

the intensity of light going through that matter is decreased by a factor of $e^{-\tau}$ (the value $e = 2.718...$ is called sometimes referred to as "Euler's number"). Your body has a high optical depth to visible light but low optical depth to, say, radio waves. The reason X-ray scans of your body work is that materials of different composition, densities, and sizes all have different optical depths so the number of X-rays that penetrate to the film will vary through different sightlines.

*A blackbody, or *thermal*, radiator has a spectrum that is entirely characterized by the temperature of the material. Theoretically, a blackbody is a perfect absorber (i.e., it does not reflect light, $\tau \to \infty$) at all wavelengths and, because it would have to be in thermal equilibrium with another blackbody attached at the same temperature, must also radiate the energy it absorbs (so as not to heat up). The peak in a blackbody spectrum is linearly proportional to the temperature of the source. People are not good blackbodies at optical wavelengths, and we know this because we reflect the ambient light in the room. But we do radiate like a blackbody near the peak of our spectrum—at mid-infrared wavelengths, around $10 \ \mu m$, dictated by the internal temperature of our bodies.

this clue is the recognition that many high-energy gamma rays (well above E_{peak}) have escaped the source despite the propensity of gamma rays to interact with other photons and produce electron-positron pairs. Such *pair production* is only possible when, in the *center-of-momentum frame* of the two photons, there is enough energy to produce two particles of the mass of an electron (m_e); that is, when the available energy is comparable to $2 \times m_e c^2 = 1022\,\text{keV} \approx 1\,\text{MeV}$.[15] The optical depth for a high-energy photon escaping from a spherical system (with an appropriate number of low-energy photons around to pair-produce with) is:

$$\tau = \sigma_T \, l \, n_\gamma, \qquad (2.1)$$

where σ_T is the cross-section for the interaction of photons with other photons to produce electron-positron pairs, l is the length of the material through which the high-energy photon travels, and n_γ is the volumetric density (units of [length^{-3}]) of low-energy photons for which the pair-production energy threshold is satisfied.* The number of suitable photons (N_γ) should be something comparable to

---

*Here, think of τ as related to the probability that a high-energy photon will pair-produce at some point while it traverses the region where other photons are present. Though light and electrons do not have physical size per se, it is instructive to think of a cross-section as an effective two-dimensional size. If the cross-section (in units of area or [length^2]) is large, then more reactions can occur because there is a higher chance of one photon interacting with another. Likewise, there will be a high optical depth if the density of interacting particles is larger or the photon traverses a longer path. An apt analogy would be trying to calculate the chance of your bumping into someone if you ran down a long school hallway with your eyes closed. That chance is low between classes when the density of other students in the hallway is low. But for the same density of people

the total energy of a pulse in gamma rays (E_{pulse}) divided by the typical suitable photon energy ($\sim m_e c^2$); the number density n_γ is just N_γ divided by the volume ($4\pi\, l^3/3$). As in §1.4, we can employ the $1/r^2$ law to estimate energy from the inferred distance (d) to the source and the fluence (S_{pulse}) of a pulse. This yields an estimate for the optical depth of:

$$\tau = \sigma_T l \left(\frac{E_{\text{pulse}}}{m_e c^2 \frac{4\pi}{3} l^3} \right) \qquad (2.2)$$

$$\approx \frac{3\sigma_T d^2 S_{\text{pulse}}}{l^2 m_e c^2}. \qquad (2.3)$$

For a cosmological GRB, we can assume a rough distance of $d \approx$ few Gpc and a typical pulse fluence of $S_{\text{pulse}} = 10^{-7}\,\text{erg}\,\text{cm}^{-2}$. Recalling the light-travel time argument from §1.4, we can infer from the variability timescale of a GRB ($\delta t \approx 0.01$ sec) that the size of the source responsible for emitting the energy in a single pulse should be of length $l \lesssim c\delta t \approx 3{,}000\,\text{km}$. Putting in the numbers, we infer that the optical depth to pair production is $\tau \approx 2 \times 10^{14}$; with a number this large, we should see no high-energy gamma rays at all! Some assumption in our calculation must have been seriously wrong—the source appears to be much too compact (small l) to allow the observed spectrum to emerge.[16]

The solution to this so-called compactness problem comes from the invocation of relativistic expansion of

---

in the hallway, τ will be higher if everyone is an oversized football player (high cross-section), and τ will be lower for a bunch of diminutive kindergardeners (low cross-section). Please do not try to reenact this thought experiment!

the radiating material. If the material is expanding at a rate close to the speed of light, then two effects work to diminish the true optical depth greatly. First, relativistic motion leads to extreme Doppler shifts of the intrinsic spectrum; a photon *observed* at tens of MeV might have been *generated* in the source at much less than a few keV. Thus, at the source, the number of photon pairs satisfying the pair-production energy threshold is greatly reduced. Second, a relativistically expanding source that is emitting can be much larger than the observed variability timescale would suggest. If the expansion velocity is v then when considering relativistic motion, it is convenient to introduce another parameter, Γ, derived from velocity called the *Lorentz factor*.* The first effect reduces the inferred τ by roughly $\Gamma^{2\beta}$, where β is the Band spectral index after the peak ($\beta \approx -3$; see §2.1.1). The second effect arises because the emitter closely lags behind the light it just emitted, so the arrival time of successive pulses is bunched up for the distant observer.[17] As will be discussed more fully in the footnote on page 62, a pulse emitted over a time $\delta t'$ (as viewed by someone traveling outward with the explosion) appears to last just $\delta t = \delta t'/(2\Gamma^2)$. Since the effective τ is reduced by $\Gamma^{2\beta-2}/2$, an optically thin GRB ($\tau < 1$) requires (for the example values we used here) $\Gamma \gtrsim 57$. This implies that the material producing the GRB must be moving at least as fast as 99.985 percent of

---

*Formally $\Gamma = 1/\sqrt{1 - v^2/c^2}$. When v is 99% the speed of light, $\Gamma = 7.09$. Likewise, $\Gamma = 100$ for $v = 0.99995 \times c$. The Lorentz factor can take on any value greater than or equal to 1 ($v = 0$). A fast supernova explosion has Lorentz factors of ~ 1.001–1.003.

the speed of light! Amazingly, detailed modeling of some specific GRBs[18] have given minimum Lorentz factors of $\Gamma > 1,000$.

Relativistic motion in other high-energy phenomena (e.g., the jets emanating from supermassive black holes) is not itself unusual, but the values of Γ inferred for GRBs are larger by an order of magnitude than any other studied phenomenon in nature. Less than two years after the 1973 discovery paper, Mal Ruderman reviewed the basic physical theories for the origins of GRBs.[19] In this review, he noted how the compactness argument pushed the inferred Lorentz factor for GRBs at cosmological distances to uncomfortably large values, thus providing weak support for a Galactic origin. The Galactic origin, with a lower distance d, would require only mildly relativistic outflow (see equation 2.3). This is a common approach in science: take an idea (e.g., a cosmological origin for GRBs) to its logical conclusion and show its extreme nature—and, perhaps, absurdity—relative to a more simple possibility (i.e., a Galactic origin). But this weak support for a Galactic origin is actually a prime example of the failed application of *Occam's Razor*. Unfortunately for both Occam and Galactic models, we have learned that GRBs do not shy away from extrema but instead appear to delight in confounding theorists and their intuition.

2.2.2 Fireballs and Internal Shocks

Well before the light of a GRB escapes, the outflowing material needs to accelerate to the relativistic speeds we infer. The basic picture (see figure 2.5) is that a significant

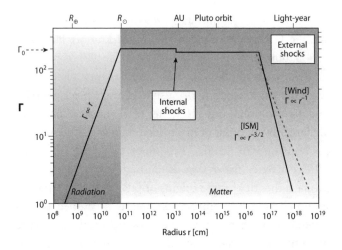

Figure 2.5. The evolution of the Lorentz factor in a GRB as a function of radius from the explosion. Energy is deposited by the central engine at a radius of $c\delta t$ (about the radius of Earth, $R_\oplus$), and Γ of the fireball increases linearly with radius until a maximum Lorentz factor Γ_0 is achieved (at about the radius of the Sun, $R_\odot$). Internal shocks occur at $r \approx c\Gamma_0^2 \delta t$, at about one astronomical unit [AU], and some energy is lost corresponding to a decrease in Γ (here we take the efficiency of conversion to gamma rays—see §3.2.1—to be $\eta = 0.1$). At a radius comparable to a fraction of a light-year, the blastwave begins to decelerate due to interactions with the external medium. It is during this external shock phase that the afterglows are produced (see §3.2).

amount of energy (which must be at least equal to the energy observed in the gamma rays) is rapidly deposited into a small region of space of size $l \approx c\delta t$. The source is compact (§2.2.1) with the energy density so high that gamma rays readily collide to make electron-positron pairs.

These pairs readily annihilate to form high-energy gamma rays. Since most of the energy is held by the photons (as opposed to the entrained matter, like electrons and protons), we call this the "radiation-dominated" phase. By construction, this soup of particles and light (referred to as a *fireball*) is opaque: few photons escape; therefore, the energy is trapped.[20] Since there is nothing to confine the fireball, it expands. During the expansion, the energy associated with the internal random motions of the moving particles and gamma rays are converted into bulk outward flow. Basic conservation arguments—that is, requiring energy and momentum to be unchanging— show that the Lorentz factor of the expanding fireball grows linearly with the radius of the fireball.

Eventually all the mass (both in the form of electrons and protons) entrained in the fireball will hold all the energy, now in the form of kinetic energy. We say that the fireball is now "matter dominated." Since the total energy[21] of a relativistic particle with mass m is $\Gamma m c^2$, if $\mathcal{E}$ is the total energy in the fireball and M_{tot} is the total mass, it follows that the fastest the material can flow is $\Gamma_0 = \mathcal{E}/(M_{tot}c^2)$. Once this "terminal Lorentz factor" (Γ_0) is achieved, the protons in the fireball are essentially moving outward on ballistic paths. We say that the fireball has now become "cold" because there is little random motion of the protons apart from their radial trajectories— if you were moving with the outskirts of the fireball, you would see almost no motion of neighboring protons toward you or away from you. The material coasts along with Γ_0 essentially unchanging over more than two orders of magnitude in radius from the explosion site. Since we have estimated that Γ of the flow around the time of the GRB must be greater than $\sim$57, we can turn this around to estimate the mass of particles in the fireball to be $M_{tot} \approx 10^{-5} M_\odot$, recalling from §1.7 that $\mathcal{E}$ is at least $\sim$10^{51} erg.

An entrained mass M_{tot} that is more than three Earth masses may seem like a lot of material, but this "dirty fireball" is a rather pristine ball of pure energy—supernova explosions, in contrast, have hundreds of thousands times more mass taking part in the bulk expansion. If there was any more mass in the fireball, then the largest Lorentz factors would be diminished, which would violate the constraints from the observed spectrum.

If the fireball was expanding out into a vacuum, there would be no easy way to turn the kinetic energy of outwardly flowing particles into the radiated energy we see. The most basic requirement is that the particles that carry the kinetic energy must be perturbed away from their outward trajectories. The simplest possibility on the path to energy release—like a car hitting a brick wall—is to have outflowing matter (mostly electrons and protons) smash into stationary material (atoms, electrons, protons, etc.) in the surrounding material. However, the density of matter in the outflow and the density of surrounding material are so low that direct interactions at the atomic level[22] are very unlikely. Like ships passing in the night, there is simply too much space for such collisions/scatterings to take place. Instead, long-range forces must connect the particles, allowing them to transfer energy and momentum to each other "collisionlessly." Magnetic fields (near the edge of the fireball) are thought to be this mitigating glue.

When energy and momentum are transferred from particles in one fast-moving region to another, a *shockwave* is generated. We think of shocks as discontinuities in the bulk properties of one region with another (such as density, temperature, and pressure). The discontinuity remains sharp (i.e., abrupt changes over a small range in distance) because the outflowing matter is moving faster than the news of the disturbance itself can move in the material upstream from the shock. For airplanes moving faster than the speed of sound, the shock gives rise to a sonic boom. For GRBs, relativistic (collisionless) shocks are the places where kinetic energy associated with outflow is transferred and that energy is radiated (we discuss radiation from shocks in §2.3.3); with GRBs, in effect, we *see* the cosmic boom.

There are two possible origins for the creation of such shocks. First, the outflowing mass may run into material around the explosion site (*circumburst medium* [CBM]), causing it to slow up; in slowing, some of the outward kinetic energy is then transformed into random motion of the particles, which, in turn, can be effective at radiating light. Second, the outflowing fireball could catch up with another outflowing shell of material and merge with that shell. Conservation of linear momentum, energy, and mass then dictate just how much the sum of the kinetic energy in the two shells could be available to radiate away. The former, referred to as the *external-shock* scenario,[23] is attractive because of the efficiency in converting bulk flow to radiated energy, but it has a serious problem explaining the near-constant width of pulses throughout the gamma-ray light curve of the GRB.

In the external-shock scenario for the prompt emission, the observed variability of the GRB comes from a clumpy circumburst medium. The latter scenario (of shells of material catching up with other shells) is known as the *internal-shock* scenario[24]—it can easily explain the complex light curves of GRBs, but the efficiency of turning outward kinetic energy into random motions of radiating particles is low. This implies that $\mathcal{E}$ must be much larger than the energy released in gamma rays. In this scenario, the observed variability comes from the diversity of energy and Lorentz factors of shells emitted at the source. The internal-shock scenario has emerged as the most likely explanation for the conversion of the kinetic energy of the outflow that gives rise to the prompt emission; the external-shock scenario is the favored mechanism for the generation of GRB afterglows.

2.3 The Central Engine

In the internal-shock scenario, the number of pulses we see is roughly equal to the number of fireballs created by the energy source at the center of the explosion, the so-called "central engine". Likewise, the duration of the GRB directly reflects the lifetime of the activity of this engine.*

---

*In a relativistically expanding system, the time of events as viewed by distant observers appears to be compressed heavily relative to an observer sitting at the center of the event; indeed, if the evolution of a GRB was a ninety-minute soccer game for fans in the stadium, those of us watching at home would see everything happen in a fraction of a second. Consider two photons, one released toward a distant observer at radius r_1 and another released when the expanding source,

We have never actually *seen* the central engines of GRBs directly, but we can infer some basic properties of them. First, the central engine must be capable of sporadically dumping 10^{48}–10^{51} erg of nearly proton-free energy into a volume comparable to that occupied by Earth. Second, some engines must live for at least tens of milliseconds (to account for the shortest bursts) and some for thousands of seconds. Last, since the total energy output in gamma rays is about the same for the majority of GRBs we see (cf. chapter 4), there must be some common properties among the engines for different observed GRBs. However, since every GRB has a different light curve, every engine must also be active in its own way.

---

traveling with velocity v, has reached a larger radius $r_1 + \delta r$. The photon released from r_1 will arrive at a time $\delta r / v - \delta r / c$ before the photon released at radius $r_1 + \delta r$. When Γ is very large, a little algebra shows that the relation $\frac{2\Gamma^2}{c} \approx \frac{1}{c-v}$ is approximately correct. Thus, the observed time difference between the pulses is

$$t_{\text{obs}} \approx \frac{\delta r}{2c\Gamma^2}. \tag{2.4}$$

Since the blastwave is moving at very nearly the speed of light, events that happen at time t as viewed by the center of the explosion ($r_1 = 0$) occur at radius $\delta r = ct$. Thus, we find that time is "compressed" for a distant observer (relative to the time measured at the explosion center): $t_{\text{obs}} \approx \frac{t}{2\Gamma^2}$. Now imagine that we have two shells of mass traveling with $\Gamma_2 > \Gamma_1$ with shell 2 emitted after a time δt as viewed by someone at the center of the explosion. Eventually shell 2 (with Γ_2) will catch up with shell 1 traveling at a slower speed. If Γ_2 is a factor of a few larger than Γ_1 then one can show that the two shells collide, producing the GRB (see §2.3.3), at a distance $\delta r = c\delta t \Gamma_1 \Gamma_2$. From equation 2.4, it is then clear that $t_{\text{obs}} \approx \delta t$. That is, even though there is strong time compression of events, the observed duration in GRBs directly reflects the time that the central engine was active.

The energy budget and volume constraints quickly whittle down the possibilities to *compact objects*, such as neutron stars (mass $\approx 1\ M_\odot$; radius $\approx 10\,\mathrm{km}$), black holes (mass $\approx 10 M_\odot$; radius[25] $\approx 30\,\mathrm{km}$), and white dwarfs (mass ≈ 0.5–$1.4\ M_\odot$; radius $\approx 3{,}000\,\mathrm{km}$); all other possible culprits (e.g., stars like the Sun) are incapable of releasing so much energy in so little space. While a variability timescale of tens of milliseconds is a natural consequence of the small sizes, the tens-of-seconds durations (recall that these durations must reflect the time that the engine is active; see footnote on page 62) are much longer than the light- or sound-crossing time for these objects. Thermonuclear explosions involving white dwarfs and neutron stars (e.g., novae and supernovae) certainly progress on longer timescales, essentially set by the time it takes for the system to become optically thin. But those explosions are very "dirty" (i.e., too much mixing of protons in the expansion), and any short-timescale variability is generally washed out in such events.

Instead, a different sort of central engine must be responsible for powering GRBs. We now take a tour of the various possible central engines, examining the expected characteristics of the resultant events.

2.3.1 Accretion-powered Events

One scenario that explains both the timescales and energies posits that the energy source ultimately comes from mass inflow into the central engine. The "best bet" scenario[26] for this form of central engine goes something like this: because of some catastrophic event, mass at large distances starts flowing toward the central compact object. Generically, inflow toward a central mass is called *accretion*. At large distances from the central source, this inflowing mass holds an appreciable gravitational *potential energy* that is then converted to kinetic energy during the flight inward. If the mass initially has even the smallest amount of motion tangential to the direction of the compact source, it will swirl inward rather than plunge directly. You should be picturing water flowing down a drain rather than hail (or meatballs) raining down toward the ground.

If there is an overall rotation to the matter before inflow begins, a disk of swirling mass will form. There will be friction within this *accretion disk*, which serves to heat up the material in the disk and speed the rate of inflow. Some of the original potential energy of the material can be tapped in two ways: (a) the fast-moving electrons can generate strong magnetic fields with high energy densities, and (b) energetic *neutrinos* can be produced in the disk and flow away from the source. These processes of energy extraction are not steady and thus naturally could explain some of the observed variability. As long as mass continues to flow inward, some of the potential energy becomes available as an energy source.

It is interesting to examine the efficiency (η) of the potential-to-kinetic energy conversion for various central engines based on accretion. This efficiency yields a formula for the maximum energy release via accretion of this type of central engine: $\mathcal{E}_{\max} = \eta M c^2$, where M is the amount of mass flowing toward the central source. To calculate η, we calculate the potential energy difference for some mass m as it moves from a very large distance r_{large} to the (effective) edge r_{edge} of the central source as a fraction of its restmass:

$$\eta = \frac{-GMm/r_{\text{large}} + GMm/r_{\text{edge}}}{mc^2} \approx \frac{GM}{r_{\text{edge}}c^2}, \quad (2.5)$$

where M is the mass of the central object and the approximate solution comes from assuming that $r_{\text{large}} \gg r_{\text{edge}}$. A solar-mass white dwarf (WD) has a radius of about 5,000 km, so $\eta_{\text{WD}} \approx 3 \times 10^{-4}$. A neutron star with mass $1 M_{\odot}$ has a radius of ~ 10 km, so $\eta_{\text{NS}} = 0.15$. A nonrotating black hole has $\eta_{\text{BH}} = 0.06$; a fast-spinning black hole can have $\eta_{\text{BH}} = 0.42$. In contrast, nuclear-fusion processes have a maximum efficiency of just $\eta = 0.007$, twenty times less than that of a neutron star. It is clear from this analysis that WD (nuclear-driven) events are at a severe disadvantage in their ability to liberate accretion-derived energy—for the same overall efficiency of conversion of restmass to observed gamma rays, WD (nuclear-driven) events would need to involve >200 (>23) more mass than that required if the central engine were a black hole or neutron star. Since the total energy liberated in just gamma rays (which is bound to be just the tip of the energetics iceberg, as we shall see later) is $E_\gamma \approx 10^{51}$ erg $\approx 10^{-3} M_{\odot} c^2$, the total mass involved must be of order at least a fraction of a solar mass.

2.3.2 Centrally Powered Events

Alternatively, if the central source is a newly forming neutron star, significant energy can be tapped from the gravitational contraction of the proto-NS itself. The heat generated from this contraction is radiated away as hot neutrinos, some of which interact with the protons and neutrons around the source and can drive a powerful wind. The significant energy associated with the rotation of the NS could also be tapped through interactions with large, nascent magnetic fields.* Assuming the NS is a perfect sphere, the rotational energy $\mathcal{E}_{rot}$ available initially is

$$\mathcal{E}_{rot} \approx \frac{4\pi^2 M_{NS} R_{NS}^2}{5 P^2}, \qquad (2.6)$$

where P is the initial spin period of the NS. This period is thought to be very high (P on the order of milliseconds). Typical numbers yield $\mathcal{E}_{rot} \approx 2.2 \times 10^{52}$ erg. Detailed calculations[27] show that magnetic fields can dissipate this energy at large distances from the central source and accelerate matter to high Lorentz factors. The resultant outflow would be highly magnetized, which could have direct observational consequences, such as highly polarized prompt emission (see §2.1.1).

2.3.3 Energy Dissipation: The Origin of the Prompt Emission

Up to here we have summarized the engines and bulk processes that *could* make enough energy available to be radiated, but we have not discussed *how* this energy is radiated. Radiation is expected when a charged particle (e.g.,

---

*The rotational energy of a spinning BH can also be extracted via magnetic fields in the so-called Blandford-Znajek process.

an electron or proton) is accelerated (or decelerated)—
generically, some of the available free energy that is used
to accelerate the particle is carried away from the system
by the emitted light.[†]

Recall that the hot fireballs of energy and plasma (ionized gas) eventually expand, converting thermal (random
internal) energy into bulk kinetic outflow at relativistic
speeds. If the energy is deposited episodically for the
duration of the GRB, each fireball will stream away with
its own bulk Lorentz factor, magnetic fields, total energy,
and entrained mass. Viewed from afar, the outflow looks
like a relativistic wind.

When (and if!) a faster-moving shell in the relativistic
wind catches up to a slower one, rather than sail right
through unimpeded, the faster shell is slowed up as it
begins to merge with the slower shell. A collisionless shock
(§2.2.2) is established between the two shells. Viewed
by the inner (faster-moving) material, the particles in
the outer shell appear to be raining inward at a velocity
comparable to the difference of the velocities of the inner
and outward moving shells.[*] Upon entering the shocked
region, these particles (electrons, protons, and neutrons)
are deflected from their radial trajectories. This deflection
provides internal thermal energy that now becomes the

[†] We also expect light to be emitted when an atom—which can temporarily act
like a microscopic battery by storing energy—relaxes to a lower energy state.
In the case of GRBs, as we now describe in this section, the former channel is
expected to dominate.

[*] Since both velocities are relativistic, a formula for relativistic velocity subtraction
must be used. The result is that the inner shell will see relativistic velocities of the
inflowing material at a Lorentz factor somewhat less the Γ of the outer shell.

primary energy deposit that can be tapped and radiated away.

In analogy to observations of nearby supernova (non-relativistic) collisionless shocks, we believe that the observations of GRBs are consistent with a population of electrons that are accelerated to relativistic speeds within these shocks. The basic idea, called "Fermi acceleration," is that charged electrons enter the shocked region and are "reflected" by magnetic fields within the shock serving to boost their kinetic energy incrementally. After multiple reflections, the electrons have gained a considerable amount of energy and can move at speeds much larger than the outward speed of the shock itself. Fermi acceleration theory posits that an ensemble of such magnetically reflected electrons will take on a powerlaw distribution in energies. Indeed, as in supernovae shocks, there is reasonable evidence from afterglow observations (see §3.2) that most of the energy in the fast-moving electrons is distributed into a powerlaw distribution of electron energies such that, for every factor of ten increase in energy, there is about a factor of 150 fewer electrons.* Though Fermi acceleration theory is attractive, there are a number of

---

*We can write the differential number density distribution of electrons at some energy E as:

$$\frac{dN}{dE} = \text{constant} \times E^{-p}, \qquad (2.7)$$

where p is the "electron spectral index" (usually determined in afterglows to be about $p \approx 2.2$ and where the constant is determined such that the integral of $E \frac{dN}{dE}$ over all energies gives the total energy in electrons). Evidence from radio observations suggests that this powerlaw distribution of electrons actually has a low-energy cutoff at around $E_{\text{lower}} \approx \Gamma m_e c^2$, where Γ is the effective Lorentz factor of the incoming shell as viewed by the faster-moving shell.

ad hoc assumptions[28]; in truth, for supernovae and GRBs alike we do not understand precisely why the electrons are accelerated into a powerlaw distribution in energy and why the precise value of that powerlaw might be different from event to event.

The total energy now imbued into these fast-moving electrons might be a few percent of what was once the total kinetic energy of the shell. However, these electrons could not, on their own, easily radiate away their potent stockpile of energy: they must either accelerate or decelerate to produce light.[29] In the presence of a magnetic field, the trajectory of moving electrons is curved, which causes them to accelerate. When the electrons are moving near the speed of light, the light emitted by bending electrons is called *synchrotron radiation*. The electrons may also decelerate by interacting with a nearby photon: the electron serves effectively to scatter the light and impart some of its kinetic energy to the incoming light. The net result in this process called *inverse Compton (IC) scattering* is that the energy of the outgoing light is higher than that of the incoming light, at the expense of zapping the electron energy. The ratio of the amount of energy dissipated by synchrotron radiation and IC scattering is related to the energy contained in the magnetic fields relative to that in the photons in the shock.

These two emission mechanisms are the leading processes by which the few percent of the required shock kinetic energy in GRBs may be radiated (both for internal and external shocks). Unfortunately, the calculation of the emergent spectrum depends on a number of effects that are difficult to determine from first principles, such as the detailed electron energy distribution and the magnetic field strengths in the shocks. Still, with reasonable prescriptions, the basic spectral properties of GRBs appear to be reasonably fit by a blastwave emitting through synchrotron and IC. The detailed diversity of spectra among different GRBs (and during a given GRB) is not a generic prediction of a larger model but instead reflects real physical differences that are not readily calculated. Of course, performing fits of the observed data to theoretical models yields direct insight into the magnetic fields, electron spectral indices, and Lorentz factors. In this respect, GRBs are wonderful laboratories to study the microphysical properties of matter and energy in extreme situations.

3

AFTERGLOWS

We can cheer ourselves up by recalling that it is more elevated to make predictions than to explain phenomena *a posteriori*.

—Sir Martin J. Rees,
Royal Astronomical Society Presidential Address
11 February 1994

Had Ray Klebesadel and his collaborators found convincing evidence for a connection of a GRB to a supernova, the search for late-time emission after a given GRB would have been sharpened and honed from the outset. Instead, the search for afterglows was generally fragmented, unfocused, and, for decades, unmotivated by specific theoretical predictions. In the context of Galactic models of GRBs, the possibility of delayed emission at X-ray wavebands was quickly hypothesized.[1] But until BeppoSAX, there was no X-ray facility that could quickly be trained on new GRB positions to search for afterglows; likewise, no convincing signatures would be seen in archival X-ray survey data.[2] While no obvious optical or radio counterpart was known, it was natural to expect—as an extrapolation of the spectrum of the prompt gamma-ray and X-ray emission itself to lower energies—detectable short-lived emission at longer wavelengths.[3] We will refer to such light as emission of a "prompt counterpart."

Though in the early 1990s the cosmological fireball model was far from established observationally, theorists began to wonder, in the context of this model, what would happen to that residual energy in the outward flow that was not carried away during the GRB. Bohdan Paczyński and James E. Rhoads at Princeton University, making an analogy to supernovae and the bright radiating remnants of ancient supernovae, suggested that the blastwave would slow down as it interacted with the surrounding gas and dust and eventually become large enough for radio light to escape.[4] Concerted searches for radio afterglows in IPN error boxes gained steam but proved unsuccessful.[5] By 1996, Peter Mészáros (Pennsylvania State University) and Martin J. Rees (Cambridge University) began developing a detailed theory of afterglows,[6] positing that long-lived emission should be observed at all wavelengths— a *panchromatic* phenomenon—as a natural consequence of synchrotron emission from a decelerating blastwave. Though no convincing afterglow had been found to date,[7] the afterglow revolution beginning the following year would quickly confirm the basic theory. In what follows, we present the panchromatic observations of GRB afterglows, and then, in §3.2, we discuss afterglow theory and its significant modifications over the years.

3.1 Phenomenology

The moment when the prompt counterpart emission ends and when the afterglow phase begins is largely subjective. It is convenient to establish the end of the prompt phase as when the gamma-ray activity appears to have ceased. This demarcation, however, builds in an obvious set of detector- and distance-dependent biases. Instead, letting theory explicitly inform our understanding of observations (often a very good idea!), we might assign the afterglow label to any emission that we believe does not come from the same physical processes that produce the GRB itself. With this notional definition, we now review the temporal and spectral properties of afterglows as seen across the electromagnetic spectrum.

3.1.1 High-Energy Afterglows

The first GRB afterglow was discovered at X-ray wavebands following GRB 970228 by the BeppoSAX team. A fleeting pulse of X-ray light during the event[8] was caught with the WFC allowing a good initial position (~3 arcminutes radius) to be extracted from ground-based analysis. Eight hours later, using more sensitive instruments, an X-ray source ~10,000 times fainter than the initial X-ray event was seen. A few days later this source had faded by more than a factor of twenty, confirming both the spatial and temporal coincidence with the GRB; Nature would have had to be unfathomably pernicious to throw at us an unrelated yet remarkable X-ray event at nearly the same time and place as a GRB. X-ray afterglows, this first being a prime exemplar, remain the primary gateway to precise (arcsecond) localizations of all new GRBs. And so the discovery of the long-lived X-ray afterglow of GRB 970228 not only opened up a new observational channel to study the physics of GRBs but also set the stage (for the next decade and beyond) of how GRBs would be followed up at all wavebands.

For the first few years following GRB 970228, X-ray afterglows were observed in the hours to days after dozens of GRBs,[9] mostly by BeppoSAX but also, on occasion, with the U.S. satellite Chandra, the European satellite XMM-Newton, and the Japanese satellite ASCA. The data were generally consistent with a powerlaw decay in time and with a powerlaw spectrum.* If we parametrize the flux in time since the trigger (t) and observed frequency (ν) as

$$f(\nu, t) \propto \nu^\beta t^\alpha, \qquad (3.1)$$

then most of the X-ray afterglows observed beyond $\sim$1 day ($\approx 10^5$ sec) were consistent with $\alpha \approx -0.8$ to -1.2 and $\beta \approx -1$ to -2.

Though the X-ray data were sparsely sampled, especially after the first few hundreds of seconds and before the first several hours, some exceptions to the powerlaw rule emerged. GRB 970508 showed some evidence for a $\sim$day-long brightening above a powerlaw extrapolation, starting about one day after the trigger. GRB 011121

*Actually, X-ray spectra almost universally show evidence for a suppression of light at low energies relative to the expectation from a powerlaw extrapolation from high energies. This rollover is usually attributed to attenuation effects ("photoelectric absorption"; see §3.1.1.1) by atoms along the line of sight (both near the GRB event and from within our own Galaxy). But in some cases, especially for the early-time observations of X-ray light where there might be some contribution from internal shocks, some of the spectral curvature could be due to the intrinsic emission spectrum of the source.

and GRB 011211 both showed bright flares at X-ray wavebands hundreds of seconds after the main activity in the gamma rays had ceased; these events also showed marginal evidence for a steepening ("break") of the X-ray powerlaw decay at later times (well after one day).[10]

Swift, by not requiring laborious ground-based analysis to search for precise X-ray positions, can train the XRT at new GRB positions in less than a minute and can continue to return to that position for as long as the afterglow remains visible. More than 95 percent of Swift GRBs have led to the detection of X-ray afterglows. As such, Swift opened up a more complete view of X-ray light curves, starting seconds after the trigger and lasting for days. The upshot was a rather startling discovery: deviations from a simple X-ray powerlaw decline are themselves the rule not the exception.[11] Indeed, only in a few cases (e.g., GRB 061007) out of hundreds is a pure powerlaw decline at X-ray wavebands observed. Instead a rather complex picture of the afterglow has arisen. Figure 3.1 shows a schematic of what is now the "canonical" X-ray light curve, exhibiting a number of distinct phases. Though there is of course a wide range of observed behavior, in general the earliest X-rays appear to be a lower-energy extension of the prompt gamma-ray emission. This is then followed by a steep decline, where the X-ray light dims by several orders of magnitude in just a few hundreds of seconds (here the α decay constant is $\lesssim -3$). After the steep decline, a shallow-plateau ($\alpha \approx -0.5$ to 0) phase begins, lasting for tens of thousands of seconds. Then, typically a few hours after the trigger, a more rapid decline sets in ($\alpha \approx -1.2$), which sometimes leads to a more rapid decline ($\alpha \approx -2$)

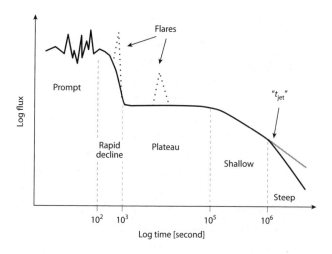

Figure 3.1. Canonical X-ray light curve following the evolution from the time of the GRB trigger. The prompt phase generally tracks the light curve behavior of the gamma rays. A short-lived period of rapid decline is then often seen, where the afterglow plummets in flux over a short period of time. A period of relatively steady flux follows, called the "plateau" phase. The "shallow" and the "steep" phases are seen after several hours. The transition ("break") from shallow to steep is often referred to as the jet break time (t_{jet}). Bright pulses seen after the prompt phase are called "flares." There are many variations on this general picture; some real light curves are shown in figure 3.2. Adapted from B. Zhang, Y. Z. Fan, J. Dyks, S. Kobayashi, P. Mészáros, D. N. Burrows, J. A. Nousek, and N. Gehrels, *ApJ* **642**, 354 (2006).

at late times.* Until Swift, only the first and last (two) phases were routinely observed, which might help explain the apparent powerlaw simplicity of the late-time pre-Swift X-ray afterglow population.

---

*This last transition to a steeply declining afterglow is usually attributed to jetting effects and will be discussed in §3.3.

Flares: Rapidly rising and falling behavior is, of course, the hallmark of what we call the prompt emission. X-ray flares/pulses during this early time are simply attributed to the same emission processes that create the gamma rays. In many cases, X-ray flaring ceases before the onset of the rapid-decline stage (at a few hundred seconds). However, many events show up to a few X-ray pulses during the rapid decline and plateau phases. The energy contained in such pulses is generally a few percent of the total energy in the prompt emission (though in extreme cases the energy can be comparable). In most X-ray flares after the prompt emission, the total duration of the flare appears to be about 10 percent of the time since the trigger—that is, the longer it has been since the trigger, the longer an X-ray flare lasts. There are some counterexamples to this rule of thumb, however: Chandra X-ray observations following the short-duration GRB 050709 showed a flare more than one day after the GRB, but this flare lasted less than a few hours.[12] We will discuss the interpretations and importance of flares in §3.2.1.2 and §5.2.3.

3.1.1.1 Spectral Features

The analog to temporal breaks are departures from power-laws in the spectral domain. Spectral features can be either in the form of absorption (less flux than we expect) or emission (more flux).

Absorption: Several GRB X-ray afterglows appear to show evidence for a rollover at low energies (see figure 3.3). This apparent absorption—which of course requires us to assume what the afterglow would have looked like without absorption—is usually attributed to suppression (commonly referred to as "attenuation") of the emergent afterglow light by the material in between us and the GRB. It manifests itself as an energy-dependent optical depth (with $\tau \propto E^{-8/3}$) and is attributed to interaction of X-ray photons with common atoms in the Universe (H, He, C, O, and so on). The degree of this interaction is measured as the "photoelectric cross-section" because when a photon of a certain energy or higher interacts with an electron in one of these atoms on its otherwise merry way to our detector, there is some chance it will instead get absorbed by that atom. Its deposited energy is then used to liberate a bound electron, causing that atom to be ionized. The normalization of the optical depth is then just proportional to the amount of absorbing atoms along the line of sight.* We can use the observed amount of absorption to infer something about the average properties of the gas and dust in the regions around the GRB and in the galaxy where it occurred.† Interestingly, the inferred amount of absorption in the host galaxies of GRBs is larger, for most events, than the degree of absorption we would expect to find if we poked random *sightlines* through the Milky Way. The implication of this will be discussed

---

*Mathematically, this can be expressed as $\tau(E) = \sum_i \sigma_i(E) n_i(l) l$, where $\sigma_i(E)$ is the photoelectric cross-section (expressed in units of [length^2]) at energy E of atomic species i, n_i is the average density of the atomic species i (expressed in units of [length^{-3}]) in a cloud of size l. The quantity $n_i l$ is the *column density* of species i and has units of [length^{-2}].

†In practice, one must first remove an inferred contribution to this absorption from our own Galaxy and also create a model for the relative contributions of different atomic species in the host galaxy, accounting for the redshift of the GRB.

further in §6.2. It is also worth noting that there have been controversial claims of a changing (both increasing and decreasing) amount of apparent absorption in a few GRB X-ray afterglows. Increasing absorption is very difficult to explain physically but could be explained away by invoking intrinsic curvature of the afterglow spectrum (as might be expected from a Band-like spectrum [§2.1.1], instead of the usually assumed intrinsic powerlaw).

Emission: In the first few seconds of a GRB, more than 10^{55} high-energy photons start streaming out from the source. If there is any material around the GRB—in the circumburst environment—it will certainly notice the deluge of these photons. Lighter elements, like hydrogen and helium, will almost certainly be completely ionized but heavier elements, like iron (Fe) and nitrogen (N), may retain some of their electrons. Some of the electrons may be pushed to an excited state by the high-energy photons but rapidly fall back to lower-energy states and, in doing so, emit lower-energy photons at specific wavelengths. This process is called "fluorescence" and is the same physical process that makes the emitted spectrum of fluorescent lights and the beautiful colors in Hubble Space Telescope images of planetary nebulae. In the BeppoSAX era there were claims of X-ray line emission in some GRB afterglows, which were interpreted as being due to large amounts of iron around the GRB, a potentially very important clue to the progenitors. Unfortunately, the evidence for line emission was never very strong, and after a few years of controversy about the statistical significance of line claims[13] coupled with the lack of significant line detections in Swift GRBs, the existence of X-ray line emission in GRB afterglows does not appear to have stood the test of time.

3.1.1.2 Gamma-Ray Afterglows?

There is, of course, nothing special about photons observed with energies within the arbitrarily defined X-ray bandpass. So the natural expectation is that afterglows should be manifest at higher and lower energies. Prompt variable emission of the GRB itself very often outshines the expected gamma-ray afterglow light, but there is good evidence from BATSE that many long-duration GRBs show evidence for low levels of smoothly decaying gamma-ray light following each event.[14] Some of the brightest events, particularly GRB 980923 and GRB 920723, show not only clear evidence for a long-lived smooth component of gamma-ray emission after the main variability has ceased but that this emission is dramatically softer than the spectrum of the prompt GRB.[15] Interestingly, long-lived emission detected by *Fermi* at the very highest energies (>100 MeV) appears to be well accommodated[16] by a simple theory for the afterglows (see §3.2.1). That is, afterglows appear to be pervasive phenomena from GeV to keV energies.

3.1.2 Ultraviolet-Optical-Infrared Afterglows

The discovery of an X-ray afterglow in GRB 970228 was followed by the discovery of an optical afterglow coincident with the initial $\sim$3 arcminute WFC localization of the prompt emission and the X-ray afterglow itself.

X-ray-afterglow discovery has become an important part of what is now a routine chain of observations: crude prompt gamma-ray localizations lead to X-ray-afterglow discoveries which lead to optical afterglow discovery.[17] However, the discovery rate of optical afterglows is not nearly as high as with X-ray afterglows: only about 50 percent of Swift GRBs have a detected afterglow at optical, infrared (IR), and/or ultraviolet (UVOIR*) wavelengths (although the majority of the nondetections can be understood as due to insufficiently sensitive searches). In §6.2 we will discuss the nature and importance of "dark bursts," a label ascribed to the ≈10–15 percent of events where no good optical afterglow was found despite exhaustive searches.[18]

At very early times, while the gamma-ray activities are in progress and just after, the UVOIR behavior is less well studied. Less than two dozen GRBs have contemporaneous measurements of long-wavelength and gamma-ray emission. For logistical reasons[†] such events tend to be the longer-duration GRBs, and, clearly, the brighter the afterglow the better the chance that small and nimble telescopes have for detecting the afterglow. The first contemporaneously detected optical afterglow was GRB 990123, found with the Robotic Optical Transient

---

*The relatively small wavelength span of UVOIR wavebands, from several hundreds of Ångström ($=10^{-8}$ cm) to a few ten thousands of Ångström makes this a natural grouping to discuss with the same physical emission and absorption processes.

[†]Relaying new positions to ground-based telescopes takes time (few seconds at best) and then getting those telescopes to start taking data also takes time (several seconds at best). The Ultraviolet-Optical Telescope (UVOT) instrument on Swift also begins taking data after the spacecraft slews, which is typically 30–60 seconds after the trigger.

Search Experiment (ROTSE). The extreme peak brightness of that afterglow (*magnitude* 9 in the visual band) was far exceeded by the fifth magnitude* afterglow of GRB 080319b.[19] The first contemporaneous near-infrared afterglow was found by my group[20] at Harvard University and UC Berkeley following an Integral/Swift event (GRB 041219b). To date, no contemporaneous UVOIR observations have been made for a short-duration GRB, most likely because the afterglows of such events are intrinsically fainter than for long-duration GRBs.

Those events with early UVOIR light curves show a wide diversity of behavior (see figure 3.2), from monotonically declining to long-lived plateaus to fast- and slow-rising "morphology."[21] When correcting for the cosmological time-dilation effect, all afterglows appear to be fading after about ten minutes in the *comoving frame*, and the intrinsically brighter the afterglow is after 400 seconds, the faster it appears to fade.[22] There have been a few suggestions that some contemporaneous optical light curves move in lock step with the prompt gamma-ray-emission light curve. However, most UVOIR light curves are not correlated with the GRB, appearing instead to follow the beat of their own drum. This suggests that the UVOIR afterglows are likely due to a different emission mechanism and/or emission site than the high-energy prompt emission.

Those GRBs with detected UVOIR afterglows provide a direct probe of the physics of the emission at late

---

*In dark skies, the human eye can see to about sixth magnitude and, with the aid of a small telescope could easily see a ninth magnitude object.

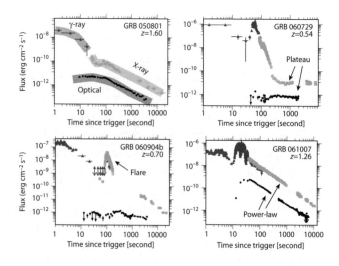

Figure 3.2. Some gamma-ray, X-ray, and optical light curves with good early optical coverage. Some of the light curve features discussed in the text are shown, such as plateaus, flares, and powerlaw declines. Adapted from E. S. Rykoff et al., *ApJ* **702**, 489 (2009).

times from the events. Just as with X-ray afterglows, the nominal theoretical expectation—that of a powerlaw spectrum decaying like a powerlaw in time—appears to be crudely borne out. However, just as with X-ray afterglows, there are important departures. Many of the best-studied afterglows show small-scale (<10%) variations about a powerlaw decline for hours to days after the trigger. Some show evidence for flaring more than thousands of seconds after the prompt emission has ceased.

There are significant departures in the spectrum of UVOIR afterglows from the nominal powerlaw expectations. There are smooth absorption features across these bands that are attributed to obscuration by dust. Dust pervades the universe (not just your living room) and is thought to be carbon- and silicon-based molecules ranging in size from a few microns to a few millimeters.[23] We believe that dust forms in the ejected material of some supernovae explosions as well as the sloughed-off envelopes of dying stars. Since it is this same "star stuff" that comprises the building blocks of stellar nurseries and stars, it is not so surprising at all that we should see its effects in very distant galaxies. Afterglow light passes through dust where its red light preferentially penetrates through the material and its blue light preferentially scatters; this is essentially the same physical interaction in the atmosphere that causes blue skies and red sunsets on Earth. So a generic property of dust is a systematic suppression of the bluest light. Figure 3.3 shows some examples of the effects of dust on afterglows. Aside from the broad dust-induced curvature, there is also a rich diversity of much more narrow absorption features in UVOIR afterglows that are due to molecules and atoms blocking certain frequencies of afterglow light. These narrow absorption lines are both telltale tracers of the redshift of a given GRB and useful in probing both the environment around the GRB and the pollution history of heavy elements throughout the Universe (chapter 4 and §6.1).

After accounting for these various roots of absorption, the intrinsic late-time spectra and light curves of UVOIR afterglows are reasonably characterized as powerlaws declining in time and frequency (see equation 3.1). Typical values[24] for the powerlaws are $\beta \approx -0.7 \pm 0.2$ and

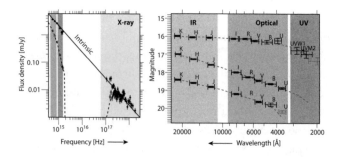

Figure 3.3. Extrinsic absorption of GRB afterglows at UVOIR and X-ray wavebands. (left) Modeled photoelectric absorption at X-ray wavebands and dust extinction at optical and UV take the observed data (lower points) to unextinguished data (higher points), showing consistency with an intrinsic smooth powerlaw spectrum (solid curve) from optical to X-ray. This GRB (070318) is thought to have one of the largest dust- and photoelectric-absorbing columns seen in GRBs. Note how the UV light (middle gray vertical region) is much more suppressed from the intrinsic powerlaw spectrum relative to the low-frequency data. Adapted from P. Schady et al., *MNRAS*, **401**, 2773 (2010). (right) Three UVOIR afterglows showing little-to-moderate amount of dust extinction. The filter names above each data point are shown. Adapted from a figure provided by D. Perley.

$\alpha \approx -1.1 \pm 0.3$. If we were to integrate this $f(\nu, t)$ from zero to infinity in time and in frequency, we would be faced with the uncomfortable inference of infinite energy release, and so it is clear that this formulation must only be relevant in a restricted range. We have already seen that for the earliest optical afterglow observations, the light curves are seen to rise (taking care of the integration problem at early times). At late times, so long as $f(t) \propto t^{\alpha_c}$ where the critical index $\alpha_c < -1$, then we also have a finite contribution to the energy release. The implication is, of course, that if $\alpha \geq -1$ during some point of the afterglow phase, the light curve must eventually bend downward. Indeed, many "breaks" are seen in GRB afterglows and we will discuss them in §3.3.

3.1.3 Long Wavelengths

At wavelengths longward of UVOIR bands lie the far-infrared, sub-millimeter, millimeter, and radio wavebands. Afterglow coverage and detection in this regime are significantly more sparse and infrequent than at UVOIR bands. However, the events that are detected at long wavelengths tend to provide some of the greatest insight into the nature of the afterglow and the burst itself. GRB 970508 was, of course, a watershed event that settled the great debate over the distance scale of GRBs (§1.6), but it also marked the first detection of radio and millimeter afterglows. The first sub-millimeter detection was with GRB 980329. Of the pre-Swift GRBs there were twenty-five radio afterglow detections (about one-third of the total localized GRBs), and a roughly comparable number were found in the first five years of the Swift era (thirty-two afterglows until January 2010). The vastly lower detection rate is attributed to the higher median redshift of Swift events and the fact that even the best radio and millimeter receivers are relatively less sensitive to afterglows than optical instruments.[25]

Unlike UVOIR afterglows that reach peak brightnesses within minutes of the GRB trigger, most radio afterglows peak only after days to weeks (see figure 3.4). There is

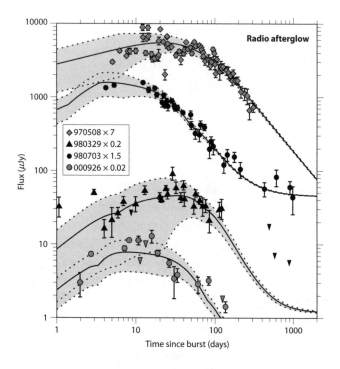

Figure 3.4. Four of some of the most extensively covered radio-afterglow light curves of GRBs discovered before the Swift era. The observed fluxes are shown along with the measurement uncertainties (vertical lines at each data point) in the fluxes; arrows represent nondetections. The solid lines show the model fits to the data (including extensive data at other wavebands not shown here; see §3.2). The gray areas represent the expected envelope of characteristic flux variability due to "interstellar scintillation." All these events took more than one week to reach maximum flux in the observed waveband of $\nu = 8.4$ GHz. The flattening of the light curves at late times may be due to the transition of the blastwave from being a relativistically expanding source to a nonrelativistic source but may also be due to a contribution from a radio-bright host galaxy; see D. A. Frail, B. D. Metzger, E. Berger, S. R. Kulkarni, and S. A. Yost, *ApJ* **600**, 828 (2004). Adapted from S. A. Yost, F. A. Harrison, R. Sari, and D. A. Frail, *ApJ* **597**, 459 (2003).

a small minority of radio afterglows that appear to peak within a few days of the GRB trigger, then decline rapidly, only to rise again on days-to-weeks timescales.

The best-studied radio afterglows appear to have erratic flickering behaviors at some observed wavebands. This flickering, first seen in GRB 970508, is attributed to so-called "interstellar scintillation" (ISS), a phenomenon also seen with other radio sources (like quasars and pulsars).[26] The key ingredients to ISS are a radio source with an inherently small size and the propagation of the radio light through a clumpy medium full of free (i.e., unbound) electrons. The electrons act as a refractory material that bends radio light, serving to amplify and dim the total intensity in different directions. Think of the bright and dark pattern on the bottom of a swimming pool on a sunny day: as you move along the bottom of the pool, just as the Earth moves through the radio-amplification pattern, you would see the light source get brighter and fainter with time. Depending on the wavelength of light, the amount and location of the "screen" of electrons in the Milky Way, and the relative distances from the Earth and the radio source to the screen, the degree of changing of amplification can be *predicted* from our knowledge of the electron distribution in the Galaxy provided we know the apparent size of the radio source. The bigger the radio source, the less scintillation we expect; essentially, the larger the apparent size of the source, the more the twinkling or scintillation effects are washed out. This is very similar to the reason planets in the Solar System do not appear to twinkle at night whereas stars at the same elevation above the horizon do.

What is so remarkable about ISS as a tool is that it allows us to make an inference of the size of the radio source even if we cannot directly resolve the source image with our equipment. In the case of GRBs, as seen in figure 3.4, it is the gradual suppression of scintillation that tells us not only that the radio source is getting bigger with time but also how physically big it would be if we could measure it directly. Thus, the consequence of the radio light curve of GRB 970508 coupled with the extragalactic distance measurement was to confirm one of the basic suppositions of GRB theory: the afterglow light was produced with a source expanding at a velocity near that of the speed of light.[27] This remarkable conclusion is reached with very little appeal to the nature and physical origin of the radio source itself. Six years later, the radio afterglow of GRB 030329 was observed at high image *resolution*, allowing a direct measurement of the expansion rate of the blastwave: more than one month after the event, the blastwave was inferred to be expanding relativistically with $\Gamma \approx 7$. These observations[28] of GRB 030329 remain the simplest and most robust evidence to date for the relativistic nature of GRB outflows.

3.2 Origin of the Emission

The Mészáros and Rees theory of afterglows[29] arrived just in time to give physical context to the discovery of the first afterglows. At its heart, the theory posits that the same blastwave that makes the prompt emission continues to expand into the ambient medium. Eventually, just as an

American football linebacker plows his way through a sea of massive bodies, the blastwave loses steam and slows up as it is bombarded by what appears to be the oncoming mass of the circumburst material.

In this deceleration, the kinetic energy tied up in the outward flow gets channeled into random motions of electrons, protons, and neutrons behind the head of the blastwave. An external (collisionless) shock is formed (§2.2.2), which we call the *forward shock*. For essentially the same physical reasons that the internal shock releases gamma-ray light (§2.3.3), this "hot" blastwave now begins to radiate (albeit at longer wavelengths than during the prompt phase). The dominant emission mechanism of afterglows is thought to be synchrotron radiation (radiating relativistic electrons in a magnetic field), just as is the apparent origin of the prompt emission. Since the characteristic energies of the relativistic electrons— which directly translate into a characteristic peak in the resultant spectrum—are connected to the bulk speed of the blastwave, the emergent spectrum and the overall evolution of the blastwave are inexorably linked.

Like all good physical theories that stand the test of time, the so-called *synchrotron blastwave theory* provides a very good framework for understanding observed afterglows, and (perhaps more importantly so) it has made predictions that have been borne out by observations. Indeed one of the first papers of the afterglow era by Ralph Wijers and collaborators[30] noted the broad agreement between the Mészáros and Rees theory and the gamma-ray, X-ray, and optical observations of GRB 970228. Yet, despite many successes, the original theory clearly fails to anticipate and explain the rich phenomenology of afterglows as observed since 1997. As such, it suffices to state here that the theory is itself an evolving one, having been both refined and added to as the data have demanded. Much of the early-time afterglow data, particularly at X-ray and UVOIR wavebands, continues to present significant challenges to the current incarnation of the theory and certainly requires extensions beyond the most simple picture. We discuss the salient features of the theory in what follows and then introduce extensions to the theory in the later sections.

3.2.1 Synchrotron Blastwaves

In the simplest treatment, we posit that the decelerating blastwave is producing afterglow light as it expands "adiabatically"—that is, we assume that the amount of energy carried away by the afterglow light is an insignificant fraction of the total blastwave energy. The attractiveness of this supposition (other than it appears to be true at late times) is that it allows us to decouple the consideration of the *dynamics* of the blastwave and the calculation of the emission. If the afterglow does carry away a significant fraction of the blastwave energy, then we would say that the afterglow is "radiative," and we would necessarily have to treat simultaneously how the blastwave evolves in time and what spectrum is emitted.

3.2.1.1 Dynamics of an Adiabatic Blastwave

On simple conservation-of-momentum grounds, in "ordinary" explosive events (like those in SNe and atom-bomb detonations), we expect significant deceleration of the outward flow when the ejected material sweeps up from the ambient medium a total mass roughly equal to that of its own mass. However, in relativistic flows like in GRBs, the ambient material of mass M_{ambient} appears* to the outflowing material to have an increased mass $M_{\mathrm{apparent,\ ambient}} = \Gamma M_{\mathrm{ambient}}$. Since the typical Lorentz factors at the time of the internal shocks is $\Gamma \sim 100$ (§2.2.1), this means that the deceleration begins when only about 1 percent the mass of the ejected material (*ejecta*), M_{ejecta}, is encountered. For a GRB with $E_\gamma = 10^{51}$ erg release, this implies† that the afterglow phase begins after about $M_{\mathrm{ambient}} = 10^{27}$ gm of material is swept up. Under the simplest assumption, we take the number density of particles n around the GRB to be constant and about equal to that of the density of hydrogen in the interstellar medium of our Galaxy; that is $n = 1$ cm^{-3}. Therefore, the mass swept up by the blastwave at radius r is just the volume times the mass per particle ($m_H = 1.674 \times 10^{-24}$ gm) times the density of particles.

---

*The apparent "relativistic energy" of a particle of restmass m moving with Lorentz factor Γ (see §2.2.1 for a definition) is $E = \Gamma mc^2$: just its restmass (mc^2) plus the additional kinetic-energy component. At low velocities v this additional component is $\frac{1}{2}mv^2$.

†Recalling that the total energy release in gamma rays is typically $E_\gamma \approx 10^{51}$ erg, it is convenient to write the kinetic energy contained in the blastwave as $E_k = \psi E_\gamma$, where ψ a constant typically equal to about ~ 10. We can write the total energy before the prompt emission as $E_{\mathrm{total}} = E_k + E_\gamma$. If the energy promptly released in gamma rays is $E_\gamma = \eta E_{\mathrm{total}}$, where $\eta \approx 0.1$ is the efficiency of conversion to gamma rays, then $\psi = 1/\eta - 1$. That is, $E_k = E_\gamma \left(\frac{1}{\eta} - 1\right)$. If the maximum Lorentz factor of the blastwave is $\Gamma_0 \approx 100$, then $M_{\mathrm{ejecta}} = E_k/(\Gamma c^2) \approx 10^{29}$ gm $= 5 \times 10^{-5} M_\odot$.

Setting this equal to $M_{\mathrm{ambient}} = E_k/(\Gamma^2 c^2)$ we have

$$\frac{E_k}{\Gamma^2 c^2} = \frac{4\pi r^3}{3} m_H n \rightarrow r = 5.2 \times 10^{16} \text{ cm} \quad (3.2)$$

Thus, for typical parameters, the deceleration begins when the source has reached about 3,500 astronomical units [AU], roughly one hundred times the size of the orbit of Pluto. Traveling at nearly the speed of light, it would take the blastwave about twenty days to reach this radius. But because of time compression for distant observers, we see this deceleration occur much more quickly by a factor of $\sim 2\Gamma^2$; for us, the blastwave appreciably decelerates in just a few minutes or less.[31] Indeed if we associate the brightest moment in the early afterglow with this deceleration time, we could use the afterglow peak time to estimate the initial Lorentz factor.[32]

It is clear from equation 3.2 that if the total energy in the blastwave remains roughly unchanged in the adiabatic phase ($E_k \approx$ constant), then the Lorentz factor must decrease with increasing radius: $\Gamma \propto r^{-3/2}$. Since the observer time is $t_{\mathrm{obs}} = r/2\Gamma^2 c$ (see footnote on page 62), this implies that the radius of the blastwave apparently changes as $r(t_{\mathrm{obs}}) \propto t_{\mathrm{obs}}^{1/4}$ and that the Lorentz factor changes as $\Gamma(t_{\mathrm{obs}}) = t^{-3/8}$. If the deceleration from $\Gamma = 100$ starts at thirty seconds, under this consideration the source becomes nonrelativistic ($\Gamma \approx 1$) after about $t_{\mathrm{obs}}(\mathrm{nonrel}) = 30 \sec \times 100^{8/3} \approx 2.5$ months.

Of course, GRBs being more complicated beasts than perhaps we would like them to be, there are some expected modifications to this simple set of derivations. The adiabatic assumption is probably a good one at late times but we certainly do expect that the afterglow carries away a nonnegligible faction of E_k during the bright early phases. This brief radiative regime causes Γ to decrease more rapidly with radius and in time relative to the adiabatic case. Likewise, it is probably too simplistic to hold that the blastwave should be plowing into a perfectly homogeneous medium on scales of AU to parsecs. Instead, if the density around the GRB event decreases with increasing distance from the explosion site (the next simplest assumption; §4.1.1), less and less mass is swept up as the blastwave expands relative to the homogeneous case. Therefore, the blastwave will decelerate more slowly with increasing radius relative to the homogeneous case. So in rederiving the relationship between Γ and r we would need to replace $n m_H$ in equation 3.2 with a functional form for the mass density as a function of radius. Figure 2.5 in the preceding chapter shows the dynamics of the blastwave during the (external shock) afterglow phase. The evolution of Γ labeled "wind" (see also §4.1) assumes that the circumburst density falls as the square of the radius from the explosion site.

3.2.1.2 Spectra from the Blastwave

We now have a reasonable prescription for the radius of the blastwave as a function of time, the Lorentz factor, and the total energy. A shock will have developed at the head of the blastwave such that the properties of matter and magnetic fields will be substantially different inside and outside the shock. To calculate the emergent spectrum, we need to know the conditions within the shock. Such conditions can be derived by assuming conservation of momentum, energy, and mass. It is beyond the scope of our treatment to carry through such calculations, but suffice it to say that the density of particles is higher in the shock as is the subsequent energy per particle. Most important is the notion that these shock properties are directly related to the bulk Lorentz factor of the blastwave at a given time. So as the Lorentz factor decreases, for instance, the density of particles behind the shock also decreases. Similarly, the average energy per particle also decreases with decreasing Lorentz factor. Both of these quantities are critical for determining the emergent spectrum.

Unfortunately, basic conservation laws do not tell us uniquely what the distribution of energies are for shocked particles, nor do we have an easy route to determine the magnetic field strengths in such shocks. Like all seasoned physicists we parameterize our ignorance, taking the distribution of shocked electrons to be a truncated powerlaw*

---

*A powerlaw distribution is not a blind guess but is informed by what is inferred through observations of SNe shocks in the Milky Way. We set the shocked electron number density (units of [electrons cm^{-3}]) to be a powerlaw distribution that is proportional to the overall number density behind the shock (itself proportional to the density in the unshocked circumburst medium and the bulk Lorentz factor of the blastwave). As first discussed in §2.3.3 in the context of the prompt emission, this distribution is set to

$$n(E) = \begin{cases} n_{\text{shock}} E^{-p} & E \geq E_{\min} \approx \Gamma m_e c^2 \\ 0 & E < E_{\min} \end{cases}, \qquad (3.3)$$

That is, we say that the electrons are shocked into a powerlaw distribution of energies (or equivalent velocities) with the minimum energy set by the Lorentz factor of the blastwave. In the case of supernovae—which we might think of as less powerful accelerators of electrons than GRBs—the value of p is inferred (see J. H. Buckley et al., $A\&A$ **329**, 639 [1998]) to be around 2 to 2.2.

and the magnetic-field strength to be a constant fraction of the instantaneous energy density in the shock. With these ingredients in hand, the emergent synchrotron spectrum can be calculated.

Instantaneous Snapshot: Figure 3.5 shows some snapshot spectra of GRB afterglows from radio to X-ray, comparing data and theory. The peak luminosity of the emergent spectrum is thought to come from radiation by the largest number of electrons: for the assumed electron distribution (equation 3.3), those that have the lowest energies (E_{min}) are the most numerous. In carrying through the calculation for peak luminosity, in the simple case of a circumburst medium that is uniform (i.e., with $n = $ constant), the dependencies on Γ of the shock cancel out, and we are left with a constant brightness for most of the relevant observing time. The frequency at which the emergent spectrum is brightest is called the "peak" (or "maximum") frequency, symbolized as ν_m. At frequencies higher than ν_m, the spectrum at a given time is influenced by the electrons at all energies. This portion of the spectrum is calculated to be a powerlaw $f(\nu) \propto \nu^{\beta}$ where $\beta = -(p-1)/2$. That is, all that matters is what the slope of the electron energy distribution happens to be. At the very largest frequencies (in the UV or X-ray band initially), the slope of the spectrum is expected to be steeper. The reason for this is that the very fastest-moving electrons are actually radiating an appreciable amount of their kinetic energy quicker than electrons with those energies can be replenished by the shock. The spectral interface between the two regions is called, appropriately enough, the *cooling break*, and the frequency at which this occurs is denoted

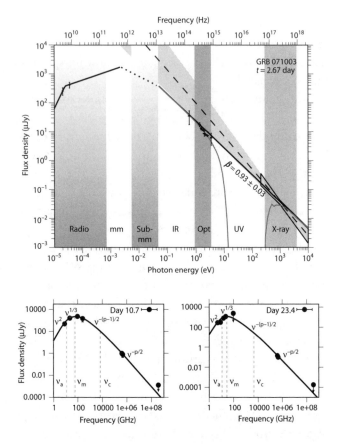

Figure 3.5. Observations and model fits to basic afterglow theory. (top) An instantaneous snapshot spectrum of the afterglow of GRB 071003 2.67 days after the GRB, showing both observed data and a synchrotron-shock model. The faint thin line that deviates from the powerlaw at UV and X-ray wavebands shows the effect of dust and photoelectric absorption (§4.1.1). The dashed line and gray wedge show the extrapolation (and uncertainty in that extrapolation) of the X-ray data to smaller frequencies. Connecting the X-ray data to the dust-corrected optical data gives an acceptable

(*Continued*)

ν_c. At frequencies less than ν_m, the spectrum is dominated by the basic low-energy synchrotron tail of lowest energy electrons. Here $\beta = 1/3$, regardless of the value of p. At the very lowest frequencies (typically at radio wavebands), the optical depth of the blastwave is greater than one, leading to a rollover in the spectrum. The characteristic frequency ν_a of this rollover is called the "synchrotron self-absorption" frequency.

Evolution in Time: Since Γ is, generally, decreasing with time, this whole snapshot spectrum should evolve in time (see figure 3.5). In the simplest case where the density of the circumburst medium is constant (see equation 3.2), the overall maximum brightness remains unchanged, but the location of this peak (ν_m) marches monotonically downward from the X-ray to the optical to the radio bands over time periods ranging from a few minutes to a few weeks. So if you were to observe at a fixed frequency—say, at optical wavelengths—you would see the source get brighter as the peak frequency starts to approach your observing band. Then, after the peak frequency

---

Figure 3.5. (*Continued*) fit of $\beta = 0.93$ (see equation 3.1). The radio data are consistent with expectations of a steepening due to synchrotron self-absorption. Adapted from D. A. Perley et al., *ApJ* **688**, 470 (2008). (bottom) Evolution of the observed spectrum in GRB 070125 with the three break frequencies labeled. Note how ν_m and ν_c appear to evolve with time. From P. Chandra et al., *ApJ* **683**, 924 (2008).

moves through your band (toward smaller frequencies), you would then see the afterglow dimming. The flux before maximum appears to rise as a powerlaw and then decay as a different powerlaw. If $f(t) \propto t^{\alpha}$ at some fixed observing band, we can calculate α under different assumptions. Importantly, after ν_m has passed through the observer's band, α is expected to be directly related to the electron distribution index p, such that $\alpha = -3(p-1)/4$.

The interesting ramification of this basic theory is that, if you can measure the spectral slope of the afterglow at some frequency, you can predict how the flux should change with time (and vice versa). Put another way, the evolution with time of the afterglow can be used, in principle, to *infer* the basic parameters of blastwave: p, Γ as a function of time, E_k, the magnetic field strength, and the density structure of the circumburst environment. Getting constraints on the macrophysical and microphysical parameters has become a small cottage industry in the GRB field. Unfortunately, however, this product line is derived from a very expensive set of raw materials: to measure uniquely and robustly the parameters of interest in a given GRB afterglow, a lot of data needs to be acquired across the electromagnetic spectrum.[33] Without good coverage, ambiguities persist about the location of the break frequencies, which lead to very different determinations of the parameters of interest.

Even in the presence of good temporal and spectral follow-up, there are several wrinkles that complicate the interpretation of the data in the context of the basic synchrotron blastwave theory:

- **Light Travel-Time Effects**: While the calculation of the instantaneous spectrum may be correct at a given radius and for a given Γ, the emitting region is (most simply) spherical. As a result of this and the finite travel time of light, photons emitted at the same time and radius reach the observer at different times: photons from the edge of the emitting region take longer to get to us. Put another way, at a given observer time, the observed true spectrum is a nontrivial admixture of light from material at a variety of radii and with different Γ factors. This tends to smooth out transitions across spectral breaks to the extent that an analytical calculation of the spectral breaks is inadequate to describe fully the true emergent spectra. Numerical simulations are required to produce a more realistic set of model expectations.

- **Other Emission Components**:
  - **Reverse Shocks**: Under certain conditions, a "reverse" shock that moves back through the ejected mass of the blastwave may have much of its energy converted into random motions, which in turn produces a bright flare of emission that can last for minutes at optical wavebands and days in the radio. GRB 990123 and GRB 021004 are classic examples of afterglows thought to be dominated at early times by bright reverse shocks in the optical. GRB 990123 also showed evidence for a bright radio flare at early times,

taken to be additional evidence of reverse shock emission.[34]

- **Inverse Compton Scattering**: If a photon produced from the synchrotron process encounters a fast-moving electron in the shock, it can be scattered to a much higher energy. Such inverse Compton scattering (see §2.3.3) will suppress the emergent flux of low-energy photons and produce an excess of high-energy photons. Since some X-ray afterglows appear brighter than they ought to be under the simple blastwave theory, IC has been successfully invoked to explain the excess.

- **Other Contributing Distributions**: Even if the emergent spectrum were entirely due to synchrotron radiation by electrons, other departures from the simple spectra would occur if there was something other than a powerlaw distribution of electrons in the shock. Moreover, since we do not have a good a priori reason to expect that the magnetic field strength should be a constant fraction of the energy density in the shock, it is possible that there is a true interplay between the parameters of the shock; this would lead to differences in the evolution of the spectra. The effects of protons and neutrons in the blastwave may significantly alter the emergent spectrum.

- **Other Absorption Sources**: The degree of dust and photoelectric absorption in afterglows is treated as a free parameter in the comparison of data to models. But dust and atoms near the burst may be significantly affected by the afterglow light itself. So these absorption properties are expected to change with time, complicating the modeling. Atoms (including barren protons from ionized hydrogen) within the blastwave may nontrivially alter the emergent spectrum.

- **Obvious Light-Curve Departures**: Flares, plateaus, rebrightenings, and repeated undulations (sometimes called "bumps and wiggles") in afterglows are not part of the basic synchrotron theory yet appear to be common among some of the best studied GRBs. Some of these effects have been modeled in the context of a long-lived central engine that replenished the blastwave with more energy well after the prompt GRB phase. Another possibility is that energy, mass, and Lorentz factors may be different in some portions of the blastwave. This is a possibility we return to in §3.3.

Some of the most active areas of research in GRBs is in understanding the relevancy of these departures from basic theory on what we observe. One of the main points of consternation is that the richness of the phenomenology of GRB afterglows has yet to be captured by a complete

theoretical framework. Not only are there many free parameters in the model components that we do know about
but also, when a certain afterglow cannot be fit well, new
components must necessarily be envisioned. Such after-
the-fact theory, even if physically motivated, tends not
to be very satisfying to most scientists. It is one thing
to have a set of useful model components to help give
physical context to what you just saw. But, more than
this, observers are especially eager for model predictions
that can be falsified with new data. Perhaps the most
worrying concern looking ahead is that the edifice of the
synchrotron shock model has become so complex that it
can only be made more complex,[35] but yet cannot be
falsified.

Despite the many complicating components and the
general concerns about complexity, some basic results
appear to be robust. First, it seems as though many of
the p values range from 2–2.5, nicely consistent with
what is inferred in nonrelativistic supernova shocks.[36]
Second, while the circumburst density appears to range
over many orders of magnitude, with $n \approx 1$ cm^{-3} (see
equation 3.2) taken as the fiducial value, one of the rather
interesting results is that, with only a few exceptions, the
density of the circumburst medium appears to be uniform
(as opposed to decreasing with increasing distance from
the explosion).[37] Last, the efficiencies of conversion from
kinetic energy to gamma-ray release appear to be about the
same ($\eta \approx 0.01$–0.1) for short and long bursts for the full
range of E_k.

3.3 Evidence for Jetting

The inference of the total energy release from what we see relies on a crucial assumption about the geometry of the explosion (see figure 3.6). In the simplest case, we assume that all the energy is released isotropically (E_{iso})—that is, uniformly in all directions. But our detectors, the ones that intercept those few high-energy photons from a distant GRB, cover only a tiny fraction of all the possible directions to where a GRB could emit light. Would a satellite at the edge of the Solar System with an identical detector pointed in the identical direction measure the same brightness as one near Earth? Probably. What about a satellite in another galaxy? Maybe not. At vastly different places in the Universe, there is no guarantee that the same GRB event would appear the same. Indeed, some places in the Universe would never see a specific GRB if no photons are emitted toward that specific direction.

Jetting (or collimation) of mass outflows appears to be a common phenomenon. Some quasars and other "active" galaxies (with massive black holes swallowing up and spitting out matter) are seen to have narrow jets that persist all the way from parsec to Megaparsec scales. Small-mass analogs in our Galaxy, called microquasars, also show jets that contain a lot of kinetic and magnetic energy. Even newly forming star systems have jet-like outflows of mass during their creation. The physical origin of jets varies across phenomena but is not fully understood in most cases. However, when rotation is involved it is clear that the axis of rotation forms two preferred directions.

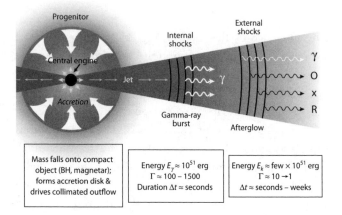

Figure 3.6. Schematic of the relevant structures, emission regions, and energetics for the internal-external shock model. The canonical energy release (using a correction for the jet opening angles) is $E_\gamma \approx 10^{51}$ erg. This is somewhat lower than the energy E_k entrained in the outflow. The inferred jet angles are generally smaller than shown; the typical opening angle of the jet is a few degrees. Adapted from P. Mészáros, *Science* **291**, 79 (2001).

If a magnetic field axis is aligned (or nearly aligned, as in the case of the Earth) with the rotation axis, then charged particles (such as electrons and protons) naturally prefer to flow to and from this axis. All this is to say that given the preponderance of jets, especially when matter is flowing near compact objects like black holes, it is natural to posit that GRB outflow could indeed be collimated.*

---

*We will return to this in the context of GRBs from massive stars in §5.1.1.

How could we infer the geometry of the explosion without being able to image it directly? It turns out that afterglow theory predicts a fairly robust behavior: a "break" in the light curve that happens at about the same time across the electromagnetic spectrum. A break is a noticeable downturn in a light curve; that is, before a break the flux is changing more slowly than after the break.* The origin of this "jet break" stems both from the geometry of the explosion—in this case, some posited collimation ("jetting") of the blastwave—and the fact that it is expanding relativistically. Basic considerations from Special Relativity show that particles moving near the speed of light appear to emit most of their energy in the direction they are heading. It is as if the faster you traveled on a dark road, the more focused your headlights would appear to become to someone watching you breeze by. Mathematically one can show that most of the energy is "relativistically Doppler beamed" into an angle of size about $\theta \approx 2/\Gamma$ (units of [radians]) when Γ is much larger than unity. So if particles in the blastwave are all moving at about $\Gamma = 100$, then anyone sitting within about one degree of the direction of motion of a given electron would see (synchrotron) light from that electron.

So as the blastwave slows down, θ increases, allowing some outside observer to see more and more of the afterglow-emitting surface. The calculation of a powerlaw decline (after ν_m has swept past the observing frequency) assumes that the observer sees more and more of a spherical

*The transition from the "shallow" to "steep" phase in figure 3.1 is an example of a light-curve break.

explosion. What if the explosion is not spherical but more of a collimated fountain of ejecta? In this case, if you happen to be looking down the barrel of this jet coming toward you, at early times you do not notice that it is not spherical (because Γ is large and you can only see light from an angle $<2/\Gamma$). But later, once Γ has decreased enough, you suddenly see that there is no material at larger off-axis angles. This is the origin of the break in the light curve, and since it occurs because there are fewer-than-expected electrons radiating toward you, the break must occur at the same time across the electromagnetic spectrum. At this break time,* the collimation angle of the jet must be comparable to $1/\Gamma$.

Just as α predicts β of the spectrum (see §3.1 and §3.2.1.2), so too does the value of α before the break time predict the decay rate $\alpha_{\text{post-break}}$ after the break. Indeed, in the simplest case, the flux after the jet break should drop more steeply than before as $f(t) \propto t^{-p}$ (i.e., $\alpha_{\text{post-break}} = -p$). GRB 990510 has a classic afterglow showing a break at optical wavebands (figure 3.7). This break is consistent with the break occurring simultaneously at other frequencies. Confirming a theoretical prediction,[38] the decay slope before the break also correctly anticipates the value of the decay slope after the break. If the collimation angle is found to be θ_j, then you can show that the true energy release in gamma rays is a factor of $\theta_j^2/2$ less than what you

---

*It turns out that this break time also corresponds to the moment when the jet begins to spread sideways (i.e., perpendicular to the direction of outflow), thus sweeping up circumburst material more quickly than before. This causes a rapid deceleration in the blastwave, further leading to the break in the light curve.

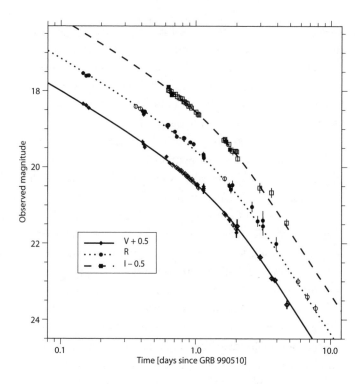

Figure 3.7. The optical afterglow light curve of GRB 990510 showing evidence for an apparent break due to jetting. Here, the break occurs around $t_{jet} \approx 1$ day following the GRB in all three colors ("V," "R," and "I"). The data in those three different colors appear to decay at the same rate before and after the break. Adapted from F. A. Harrison et al., *ApJ* **523**, L121 (1999).

would have inferred if the energy was emitted isotropically. Since, when they are measured well, typical θ_j values are in the range of a few to several degrees, this means that the true energy release in gamma rays can be 0.1–1 percent of the "isotropic-equivalent" value. Canonically,[39] it seems that this geometry-corrected value is about $E_\gamma = 10^{51}$ erg, although there are many events with lower inferred E_γ.

Before Swift there were about ten GRB afterglows that showed some indication of a break both at optical wavebands and at either X-ray or radio wavebands. However, despite (or, perhaps, because of) the huge improvement in X-ray coverage (especially within a day after the trigger), only a small minority of Swift GRBs (<10%) show the classic signs of a jet break.[†] When a break attributable to a jet is not seen (such as in GRB 061007; figure 3.2), we can use the nondetection of a break to set a lower limit on the jet collimation angle: this can be useful because we then learn that the energy release must be at least some value. Still, there are ways to wiggle out of specific collimation constraints. First, the inference of a collimation angle requires assumptions about the gamma-ray-conversion efficiency η and the circumburst density profile; for a nondetection of a jet break, a lower circumburst density would tend to push the lower limit on the inferred collimation angles to smaller values. Second, we usually infer jet collimation from late-time afterglows, but it is possible that the GRB was emitted from a more collimated jet embedded in a wider jet responsible for the late-time afterglow. The possibility of multiple jets (or more generally, a jet with differences in energy and ejecta mass as a function of position) adds yet another layer of complexity to an already complicated modeling effort.

[†]To be clear, there are *plenty* of breaks seen in GRB afterglows, but most do not conform to the temporal and spectral expectations of a jet. The physical origins of such breaks remain ambiguous.

3.4 Late-Time Observations

Given the richness of both the observations and theory of afterglows, it is fair to say that the afterparty may be more interesting than the main event. But after the afterglow from the relativistic outflow has faded and after jet breaks have kicked in, the GRB story does not end. In many cases, especially for long-duration GRBs, a faint galaxy consistent with the afterglow position (once outshown by the afterglow) is revealed. We will discuss the galaxy connection in detail in §4.2. In a few cases, well after the optical afterglow has vanished, evidence for a flattening of the radio light curve (see figure 3.4) is seen on timescales of months to years. By this time, any jet collimation has been washed out, and we can now consider the blastwave to be spherical, not much different than late-time radio emission from supernova remnants. Appealing to general arguments about the distribution of energy in electrons and magnetic fields, one can determine the radius of and the energy in the late-time blastwave. As inferred from collimation-corrected prompt energies, these few radio observations show that there is about $E_k = 10^{51}$ erg in the kinetic energy of the blastwave.[40] In the case of the nearby GRB 980329, there was a rough consistency between the blastwave radius measured from high-resolution radio imaging and that inferred from general arguments.[41]

At optical wavebands, a few dozen GRBs have shown some evidence for rebrightening effects in the late-time afterglows on timescales of weeks to months. Early suggestions that these "bumps" had the color and light-curve characteristics of a supernova were vindicated with spectroscopic observations, starting in 2003, of some bumps seen in nearby GRBs.[42] The nature and pervasiveness of these GRB-supernovae (GRB-SNe) and the implications for the progenitors are discussed in detail in §5.1. But suffice it to say, late-time observations of GRB-SNe provide a remarkable glimpse into the origin of at least some GRBs.

4

THE EVENTS IN CONTEXT

*L'accent du pays où l'on est né demeure dans l'esprit et dans le coeur,
comme dans le langage.*
[The accent of one's birthplace remains in the mind and in the heart
as much as in one's speech.]

—François duc de La Rochefoucauld,
Maxim 342, *Réflexions ou sentences et maximes morales*
(Reflections; or Sentences and Moral Maxims), 1664

Prompt and afterglow emission of GRBs are largely driven by the central engine behavior, the explosion properties, and the physics of relativistic shocks. Those seconds, minutes, and days after the main event tell a remarkable story about how the progenitors of GRBs end their life. But it is the context—where GRBs occur inside and out of galaxies and throughout cosmic time—that tell us how the progenitors lived. Indeed, we treat GRB locations like a crime scene, extracting forensic evidence to make a case about the lifecycle of a GRB progenitor; that progenitor, while on the other side of the Universe, would one day get a bunch of astronomers on Earth scurrying around to make sense of it all.

4.1 Local Scales

4.1.1 Circumburst Environments

X-ray absorption and emission lines (§3.1.1.1) in long-duration GRBs were suggested to arise from a very specific (and somewhat contrived) geometry of the circumburst medium (CBM). Had these spectral features been found to be real, a dense patch of gas-phase metals at 10^{17}–10^{18} cm from the burst would have been required. This, in turn, would have given a rather specific view of the history of the amount of material expelled ("mass loss") by the progenitor. Instead, since the significance of the X-ray diagnostics has remained low, we are forced to turn to other approaches for inferences about the CBM. We saw in §3.2.1.2 that, modeling caveats aside, GRB afterglows can be used to infer both the density of the CBM as well as its change with distance r from the explosion site. We can parameterize the density of the circumburst medium as $\rho_{CBM} = C \times r^s$, with C as some constant. All other parameters being equal, the larger the value of C, the brighter the afterglow will be. As we have seen, the values of C and s also affect the dynamics of the blastwave. The smaller the value of s, the longer the blastwave will travel at relativistic speeds. A homogeneous (uniform) medium has $s = 0$ and $C = n \times m_H$. This is the nominal expectation of a burst that occurs far from the influence of stars, in the *interstellar medium* (ISM)[1] or in the *intergalactic medium* (IGM). Typical values for the ISM and IGM are, respectively, $n_{ISM} = 1$ cm^{-3} and $n_{IGM} = 10^{-6}$ cm^{-3}.

If instead the progenitor is a dying star, we expect a different value of C and s. The simplest assumption is that this star should shed its outer envelope as a stellar wind, much like the Sun does but with more mass and at faster velocity. Assuming a constant velocity and constant mass-loss rate,* one can show that $s = -2$. Such a "windblown" CBM (§3.2.1) is the most obvious expectation for a GRB that originates from a massive-star explosion, especially one that also produces an SN: massive stars are observed to have very strong winds giving rise to $\rho \propto r^{-2}$ density profiles in massive-star SNe observed in the local universe. Yet, with model fits to observations, a constant density medium ($s = 0$) is preferred in the overwhelming majority of long-duration GRB afterglows.† For the few short-duration GRBs that have had enough data to model C and s values, there appears to be broad consistency with a homogeneous $s = 0$ medium. In many cases, there is an indication that C is significantly less than seen in long-duration GRBs; the implications of this will be discussed in §5.2.

Since the majority of long-duration GRBs are thought to come from the death of massive stars (§5.1), the $s = 0$ environment is a considerable bugaboo. Least favored, but certainly not out of the question, is the notion that our physical model of the afterglows is really incorrect and that we actually are not measuring s at all. Another possibility is

---

*This mass-loss rate is usually given in units of [Solar mass ($M_\odot$) per year] and has typical values for the progenitors of interest of $10^{-4} M_\odot$ yr^{-1}.

†The exceptions are few enough to count on one hand. GRB 011121, which had strong evidence for a supernova bump, showed reasonably strong evidence for an $s = -2$ environment. See P. A. Price et al., *ApJ* **572**, L51 (2002); R. A. Chevalier, Z. Li, and C. Fransson, *ApJ* **606**, 369 (2004).

that GRB progenitors are not like the sort of massive stars we observe in the local universe; still, it would be a highly contrived situation if the massive stars all had special mass-loss histories that mimicked an $s \approx 0$ environment. The current consensus is that the environments around many GRBs are homogeneous but that this homogeneity derives from an interplay between the outflowing wind of the progenitor that has interacted strongly with the interstellar medium and/or the remains of the star-forming region that harbors the progenitor. This idea also has its problems: the $s = 0$ medium appears to exist so uncomfortably close to the explosion site (at least as close as 10^{16} cm, where the afterglow starts radiating) that the external medium would need to be very dense ($>10^4$ cm^{-3}) to create such a homogeneous bubble. The CBM homogeneity issue, without a truly satisfying explanation, persists.

Another useful vista into the circumburst environment is to look for time-dependent changes that are attributable to interactions of the afterglow light with the material around the GRB event. Here the composition of the circumburst medium is important (not just the density). In the case where there are just atoms in the CBM, the atoms that are in our line of sight to the afterglow will absorb the afterglow light (§3.1.2) at specific wavelengths. These wavelengths (and depth of absorption) depend on the specific population of electrons bound to the atoms and the quantum-mechanical rules that dictate how those electrons can gain energy by absorbing an afterglow photon. Near the GRB (say within 100 pc), these atoms are bombarded by the deluge of afterglow photons at such an appreciable rate that one can show that, within a few seconds, essentially all the electrons of all the atoms are stripped off, leaving a fully ionized plasma of nuclei (protons + neutrons) and free electrons. If there are any simple molecules in the CBM (such as H_2 or carbon monoxide), then these too should be shredded by the afterglow light. Dust particles are complex molecules (or "grains"[2]), but these too can be destroyed by a number of photon-interaction processes. The GRB perpetrator is rather adept at wiping clean the crime scene.

Dust destruction almost certainly happens around a GRB, but no unambiguous case for dust destruction has been seen in the time variability of an early afterglow.[3] Aside from a high-ionization line of nitrogen, there are no atomic absorption lines that are attributable to atoms in the CBM—again, not that the atoms are not there, just that they are fully ionized before we can acquire a spectrum of them.[4] Also consistent with the general picture is that most GRB afterglows do not show evidence for molecular absorption, despite the fact that the star-forming regions in the Milky Way do harbor simple molecules. The notable exception is the long-duration GRB 080607, which showed significant absorption due to H_2 and carbon monoxide.[5] What set this event apart from most others is that there was a considerable amount of dust along the line of sight—perhaps the largest amount ever seen in a GRB—and that the afterglow was intrinsically one of the brightest ever inferred. This, combined with the fact that we were able to observe the event at moderately high spectral resolution quickly (within about ten minutes) on a large-aperture telescope, allowed the molecules to be observed in absorption. Though the cloud housing the molecules was reckoned[6] to be more than 250 pc from the GRB site, these afterglow observations of GRB 080607 provide the best evidence to date for a nearby environment similar to star-forming molecular clouds studied in the Milky Way.

4.1.2 Subgalactic Scales

From a progenitor perspective, the lack of telltale diagnostics of the CBM in all but a few cases is disappointing but entirely reasonable. To be sure, there is significant absorption (both broadband and in narrow lines) in long-duration GRB spectra that is attributable to the absorption nearer to the GRB than to us. However, aside from the few cases mentioned in §4.1.1, the absorption sites are thought to be in physically disconnected regions in the host galaxy (that is, far away from the GRB). GRB afterglow light pierces through random sightlines in its host galaxy, so if it encounters a pocket of dust or gas, those distinct places within the host galaxy will leave their distinct imprints on the afterglow. The only difference with CBM regions is that, because of the distance from the GRB, there are not enough afterglow photons to do any destructive (ionizing) damage. In a few well-observed cases of long-duration GRBs,[7] we have been able to witness the changes of atomic states as a function of time. In such events—observed in the few minutes to hours following the GRB—there are enough photons to *alter* the states of some ions at hundreds to thousands of parsecs from the GRB site. After the afterglow fades, some atoms relax back to their original states. By observing the rate of depopulation and repopulation of some states, one can infer the distance of the absorbing cloud from the GRB and, in addition, some basic properties of the cloud (such as density). When such data can be obtained, we get a delicious upfront and personal view of random clouds in distant galaxies in a way not possible by other means. For astronomers, it is like anonymously following the Twitter stream of a random person in a far-away country for long enough to learn about his/her community and what makes him/her tick.[8]

The vast majority of long-duration GRB afterglows observed spectroscopically do not show time-variable behavior in the absorption lines. Yet at high spectral resolution, it is clear that persistent absorption lines arise from the light of the afterglow passing through distinct clouds in the host galaxy.* We can separate the clouds by looking at slightly different wavelengths. Motion within those distant galaxies—individual clouds having slightly different speeds moving toward and away from us—gives rise to distinct Doppler shifts such that their atoms appear to absorb at slightly different wavelengths. In each cloud, we can see the relative amounts of atoms in different states of ionization. In many cases we can also determine the total amount of hydrogen in the host galaxy.†

---

*Recall from §1.6 that this is precisely how we can obtain the spectroscopic redshift of a GRB.
†What we actually infer is the column density of neutral hydrogen (N_H, units of $[cm^{-2}]$). This is done by analyzing the absorption spectrum near the dominant electronic transition of a neutral hydrogen atom—the "Lyman α line" at wavelength $\lambda = 1216$ Å as measured in the laboratory. This is the absorption that causes a neutral hydrogen atom to become excited from the ground state

Measuring the amount of metals* relative to the amount of neutral hydrogen gives us a sense of how much of the ISM of that galaxy has been enriched by synthesized (heavy) elements. We usually compare this measurement to the same measurement made in the spectrum of the Sun, deriving a value called *metallicity*. GRB afterglows have allowed us to make dozens of metallicity measurements in GRB hosts. In general, the metallicities inferred are less than that in the Sun. In the very distant universe this is not entirely unexpected, given that the GRBs are formed from stars that themselves have formed before significant amounts of metals have been spewed out from SNe.[9] One very interesting result is that, in the few GRBs that have occurred "nearby" (within one billion light-years from Earth), the metallicities inferred for those GRB hosts (and, in particular, at the very location within the host where the GRB occurred) tends to be much lower than seen in a typical nearby galaxy.[10] This is taken as evidence that the progenitors of long-duration GRBs prefer low metallicities for their formation. No absorption spectrum of an unassailably classified short-duration GRB has been acquired to date. The utility of measuring metallicities in GRBs, apart from inferring something about the progenitors, is discussed in §6.1.

(the most energetically favorable configuration) to the next excited state, which corresponds to the $n = 1 \rightarrow 2$ electronic transition (for those that have some background in chemistry or atomic physics). We will return to this in §6.3.

*"Metals," for astronomers, generally refers to all elements with more protons than hydrogen (number of protons = 1) and helium (number of protons = 2); this even includes elements we do not normally think of as metallic on Earth (such as carbon and neon).

4.2 Galactic Scales

Bright afterglows are wonderful for probing gas and dust in absorption: we need a lot of light at all wavelengths to be sure which photons are missing at which wavelengths.[11] But since afterglows can be millions of times brighter than entire galaxies until they fade, trying to capture an image of the large-scale environment around a GRB is as futile as snapping a picture of a firefly in front of stadium lights.

When the afterglows do fade, we can determine where within (or outside) galaxies the events occur. In the special cases of the most nearby GRBs, the galaxies are big enough on the sky that we can image and study the specific locations of the events with imaging resolutions* of tens of parsecs, comparable to the size of large star-forming regions. In the case of long-duration GRB 980425 and SN 1998bw, the event occurred within a cluster of stars in the spiral arm of an otherwise normal-looking galaxy. This cluster was less than 1 kpc away (but physically distinct) from the most copious factory of massive stars in that galaxy. A few other low-redshift GRBs (e.g., 060218 and 060505) also appeared to be associated spatially with ongoing and vigorous star-forming regions in the parent

---

*Resolution $\mathcal{R}$ is a measure of the blurriness of an image and is usually given in angular units ([radians] or [arcsec]) with smaller $\mathcal{R}$ being more desirable. Knowing the distance D to the GRB, we can translate $\mathcal{R}$ into an effective size scale S at the location of the GRB: $S = D \times \mathcal{R}$. The theoretical best $\mathcal{R}$ that a telescope of diameter d observing at wavelength λ can achieve is $\mathcal{R} = 1.22\lambda/d$ (other effects, such as from the atmosphere, cause $\mathcal{R}$ to be higher). Taking a large telescope with $d = 10$ m observing at infrared wavelengths (say $\lambda = 1.2\,\mu$m), $\mathcal{R} \approx 30$ milliarcsec. At the distance of GRB 060218 ($D \approx 135$ Mpc), $S = 20$ pc.

galaxy.[12] However, for the overwhelming majority of GRBs (both of the long and short varieties), the GRBs are far enough away from Earth that even the sharpest imager (such as the Hubble Space Telescope) cannot resolve the subgalactic structures to any less than a few kiloparsecs. We are, therefore, left to study the locations of GRBs in and around galaxies and the aggregate properties of those associated galaxies.

4.2.1 Locations

Less than one year into the afterglow revolution, the few long-duration GRBs that had been well localized via optical and radio afterglows proved very important for revealing which sort of neighborhoods they preferred and which sort of neighborhoods they avoided. Precise localizations[13] told us that GRBs were not coming from the centers of galaxies, as one might have expected if they were due to activity around the central massive black hole found in most galaxies. Nor were these long-duration GRBs far from their apparent hosts, in seemingly empty space. Instead long-duration GRBs preferentially occurred embedded in the light of distant galaxies. The quantitative connection with galaxy light—and particularly blue galaxy light—in many individual cases could not be shown precisely, but larger samples showed a strong statistical connection with the ensemble. What location studies of long-duration GRBs did for the community was rule out some progenitor models, leaving models closely connected to the life and death of stars as the most viable.

Once Swift began localizing short-duration GRBs, locations became similarly useful in narrowing down the progenitor culprits of that class. But, since short-duration afterglows are generally fainter than those from long-duration events, the low success rate of measuring precise locations (with optical or radio afterglow) hampers the ability to make definitive statements about associations with host galaxies.[14] Still, some short-duration bursts appear very clearly connected to the light of individual galaxies (which themselves tend to be blue, like the hosts of long-duration GRBs), and some appear to occur at large distances from their apparent hosts. There is good evidence that short-duration GRBs are more diffusely positioned about galaxy light than long-duration GRBs (which appear concentrated with the light of their hosts). However, the precise radial distribution of short bursts around their hosts is a difficult and uncertain practice (§5.2.3).

4.2.2 Host Properties

In almost all long-duration GRBs, the host associations are unambiguous, so we have a good deal of assurance that our study of those galaxies reflects the true population of GRB hosts. Long-duration hosts tend to be some of the bluest and faintest galaxies observed in distant galaxy studies. With only a few exceptions, they are also smaller in size than the Milky Way. The shapes ("morphology") of those well-imaged hosts do not look like typical spiral galaxies (like Andromeda or the Milky Way) nor egg-shaped elliptical galaxies (like M87). Instead, the hosts of long-duration

GRBs are usually characterized as being "irregular," without clear morphological structure and, in many cases, without even an obvious center (see figure 4.1). The size encapsulating 95 percent of the star light is typically a few kiloparsecs and always less than 10 kpc.* The closest analogs to GRB hosts we have in the local universe are dwarf satellites galaxies of the Milky Way, the Large and Small Magellanic Clouds.

Since the aggregate spectrum[15] of newly forming star clusters is blue (as opposed to red for a group of old stars), the colors and brightnesses paint an important picture of GRB factories: at the time of the GRB, such birth sites are producing copious amounts of stars and yet have not produced an appreciable number of stars in the past. With high-quality images of the host galaxies in a number of optical and infrared bandpasses, the *spectral-energy distribution* (SED), coupled with models of how stars evolve with time, can be used to infer a total mass of the galaxy in stars. In accordance with their small sizes, the inferred masses are all less than that of the Milky Way and comparable to that of the Magellanic Clouds.

Spectroscopically, we can study in detail the emission lines from the hosts. Emission lines, like absorption lines, are due to specific transitions within atoms and can be used as a sensitive diagnostic of the conditions of star formation. Though there is a big range, we infer that star formation in GRB hosts is comparable to that in the Milky Way;[†] but, since GRB hosts are ten-to-one

---

*By contrast, the Milky Way has a size of more than twenty kiloparsecs.
[†]The entire Milky Way is producing about 1 $M_\odot$ in new star mass every year.

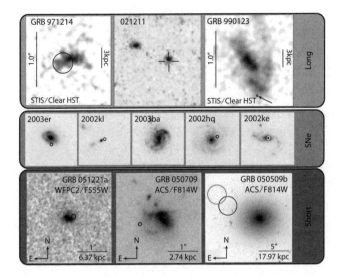

Figure 4.1. Gallery of GRB and SN host galaxies as observed by the Hubble Space Telescope. (top) Images around the locations of three long-duration GRBs with the position (and uncertainty) noted with crosses and circles. (middle) Images of host galaxies of core-collapsed SNe (not associated with any known GRB) at $z = 1$. At bottom, the hosts of three short-duration GRBs. Note that while core-collapsed SNe appear (in general) to occur in regular-shaped spirals, long GRBs are in more irregularly shaped galaxies. Some short bursts are associated with red, egg-shaped galaxies called "ellipticals," and some are associated with blue irregular galaxies. From J. S. Bloom, S. R. Kulkarni, and S. G. Djorgovski, *AJ* **123**, 1111 (2002); W. Fong, E. Berger, and D. B. Fox, *ApJ* **708**, 9 (2010); A. S. Fruchter et al., *Nature* **441**, 463 (2006).

hundred times less massive than the Milky Way, they appear to be very *efficient* producers of new stars given their relatively puny sizes. The small size and blue colors of GRB hosts suggest low metallicities (less than Solar

abundances); this is in accordance with the low average*
metallicities in GRB hosts that are inferred using emission-
line diagnostics. Interestingly, in the local universe, the
metallicities of long-duration GRB hosts are systematically
lower than in those hosts of observed SNe, suggesting
some preference of GRB progenitors to be formed in low-
metallicity environments. At higher redshifts—above $z > 1$
when the average metallicity in the Universe was lower—
the metallicities of long-duration GRB hosts inferred from
absorption-line spectroscopy is similar to the metallicity
inferred in the generic population of galaxies. Overall it
seems that long-duration GRB progenitors are less likely
to be formed at high (i.e., roughly Solar) metallicity. We
discuss some interpretations of this conclusion in §6.1.

Since star formation tends to occur in fits and starts,
with typical star-forming episodes lasting 10–100 million
years, the high star formation per unit mass in small
galaxies serves as strong (albeit anecdotal) evidence for
the connection of long-duration GRBs to *ongoing* star
formation: if long-duration GRBs were due to events
related to old stars, there would be no reason for them
preferentially to favor galaxies with ongoing star formation.
In that progenitor scenario, we would expect to see GRBs
occurring preferentially where most of the older stars
reside, in large spirals and elliptical galaxies. The definitive
establishment of the massive-star origin of long-duration
GRBs in 2003 (through a spectroscopic supernova observation following long-duration GRB 030329; §5.1.2)

---

*As opposed to *at the GRB site*, which is generally impossible to resolve except
for the most nearby GRBs (see discussion in §4.2).

confirmed the indirect evidence for a massive-star origin that was suggested by the host galaxy observations.

After GRB 030329, discussion and speculation shifted toward the progenitors of short-duration GRBs and how the progenitors of that class would be manifest in host galaxy observations. Degenerate merger models (e.g., NS–NS coalescence; see §5.2) were certainly still viable, and, indeed, the short timescales matched more closely the theoretical notion that the timescale for mass inflow following degenerate merger would, too, be short (<few seconds). When the location of short-duration GRB 050509b turned up near a massive elliptical galaxy in a cluster of galaxies, the notion that short bursts could be connected to an older stellar population than long bursts appeared to be confirmed.[16] While 10–20 percent of short bursts do appear to be associated with older galaxies, many events classified as "short" appear to be connected with galaxies similar to that of long-duration GRB hosts. Such an admixture of host types does not preclude a degenerate-binary merger scenario, nor does it force us into accepting a multiprogenitor population for short bursts. Instead there appears to be (from the modeling standpoint) reasonably good evidence for a broad distribution of merger times after star formation. So a diverse connection to star formation is consistent with, but does not require, a degenerate-binary merger scenario.

4.3 Universal Scales

Pinpointing GRBs on the sky and in redshift not only tells us *where* they happen but *when* they occur during the long history of the Universe. Since we have a good understanding of expansion history of the Universe (more on that in §6.5), there is a one-to-one mapping between observed redshift, inferred distance, and time of the event since the Big Bang. For example,[17] a distance* of 100 Mpc corresponds to a redshift of about $z = 0.023$. A redshift of $z = 2$ corresponds to a distance of 15.7 Gpc. A GRB that occurs at a redshift of $z = 5$ is at a distance of 47.6 Gpc, occurred 1.2 billion years after the Big Bang, and its light took 12.5 billion years to reach us. With over two hundred GRB redshifts now measured, we have a snapshot of the existence, activity, and prowess of objects making GRBs throughout universal time.

Figure 4.2 shows the observed redshift distribution of GRBs before the Swift-era and during. Before Swift launched, the highest (secure) redshift of a GRB was $z = 4.5$. By July 2010, the redshift record was $z = 8.2$. GRB 980425 continues to bracket the low end of the redshift distribution. The immediate implication of the redshift distribution is that the progenitors of GRBs are forming both in the early universe ($z = 8.2$ is only 630 million years after the Big Bang) and at the late epoch of today. Given the fuzziness of the long-short duration divide (§5.4) and the difficulties of definitively identifying

*There are actually a few different ways to represent distance when talking about cosmological scales, stemming from the expansion of the Universe and General Relativity. Unless noted, the distance used throughout is the so-called *luminosity distance*, which is the distance that satisfies the $1/r^2$ law for brightness as a function of distance from a source. Another important distance measurement is the *angular diameter distance* that satisfies the notion that the apparent size of objects decreases inversely proportional to distance. At small redshift, all distances are nearly identical, but they diverge at large redshift.

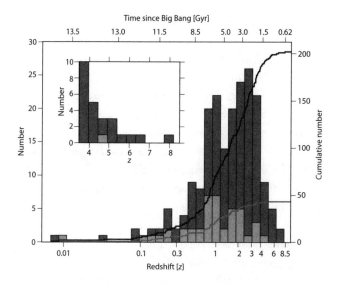

Figure 4.2. The distribution of known GRB redshifts as of January 2010. Only shown are those events where redshift was measured spectroscopically, from absorption lines in the afterglow or using emission lines in the putative host galaxy. Histograms in dark shading show all measured events, with equal spacing in logarithmic interval of redshift. Those in light shading show the distribution as it was before the launch of Swift. The solid lines show the cumulative distributions of the two histograms. Inset is the distribution above $z = 3.5$ (with redshift bins linearly spaced). The time since the Big Bang is shown at top.

a host galaxy for short events (§4.2.1), the highest redshift observed of a bona fide short burst remains somewhat uncertain.[18] There is no compelling evidence to date that the short-duration population has an intrinsically different redshift distribution from long bursts. The observed redshift distribution, which is dominated by long-duration bursts, shows a rapid rise in the number of "events per redshift interval" toward $z = 1$ and then a rapid drop after $z \approx 4$. This is similar, at least qualitatively, to the inferred rate of star formation in the Universe, which peaked from about 2 to 5 Gyr after the Big Bang.

Of course, the *observed* redshift distribution is a biased view of the *true* distribution of GRBs throughout cosmic time. There are both intrinsic effects in GRBs as well as our detection biases that confound the measurement of the GRB rate:

- **Field of View**: All satellites are blind to some fraction of the sky at any point in time. For instance, Swift is sensitive to GRBs from only about one-twelfth of the sky at any time, so the rate of GRBs must be substantially higher than detected (= about one hundred per year).

- **Detector Sensitivity and Luminosity Functions**: The total energy output (or luminosity) of GRBs is not identical: it is drawn from some distribution spanning many orders of magnitude. For any given energy or luminosity, we can only see that event to a limited distance (the faintest events can be seen to smaller distances than brighter events), and since volume increases rapidly with distance,[19] the rate of intrinsically brighter events (relative to fainter events) tends to grow with redshift. These intrinsic energy and luminosity distributions, which we do not have a good handle on theoretically, act as redshift-dependent selection functions for what we can observe. Figure 4.3 shows this rather clearly.

- **Redshift Discovery**: Not all GRB detections lead to a secure redshift measurement: for an absorption-line redshift to be made, an optical afterglow must be found, and it must be bright enough for a high-quality spectrum to be obtained (generally hours after the GRB itself).[20] Since redshifts are determined with spectroscopic lines, measuring precise redshifts becomes difficult when the most common lines fall outside the range of sensitivity of an instrument. For instance, a common set of absorption lines from singly ionized magnesium resides (in the laboratory) at ultraviolet wavelengths 2,796 Å and 2,803 Å. Between a GRB redshift of $z \sim 0.4$ and 2.2 these lines can be readily seen in an optical spectrum.[21] At lower redshift there are no common and strong lines to help us find an absorption redshift. At redshifts beyond $z \sim 3$, absorption lines due to hydrogen are the dominant features used to identify redshift. Similarly, *emission* from singly ionized oxygen produces strong and blue lines around 3,727 Å, so for GRBs beyond $z \sim 1.3$ these and many other strong emission lines cannot be detected in optical spectra. Beyond $z \approx 2$, emission from neutral hydrogen shows up in the optical bandpass, but such emission lines become fainter and fainter with increasing redshift.

- **Jetting**: Collimation and relativistic Doppler beaming (§3.3) implies that we are missing a

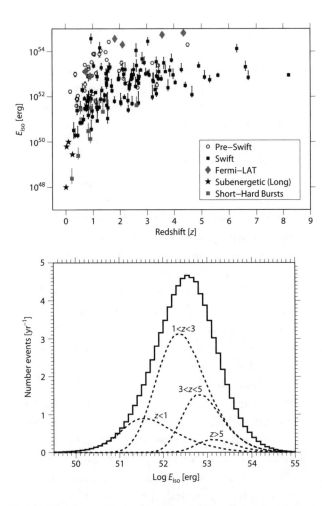

Figure 4.3. (top) The observed distribution of "isotropic-equivalent" energy release ($E_{\rm iso}$; §3.3) versus redshift for all bursts with spectroscopic redshift confirmation. Note how at low redshift a large spread of $E_{\rm iso}$ is seen, but at higher redshift only the highest $E_{\rm iso}$ events are detected. Also, the observed $E_{\rm iso}$ distribution has skewed to fainter events in the Swift era thanks

(*Continued*)

substantial number of GRBs that happen to be aimed away from us. Since we cannot easily measure the distribution of collimation angles for the GRB population, the fraction of events we are missing is relatively unconstrained—typical numbers for the beaming-corrected long-duration rates range from fifty-to-five hundred times the uncorrected rate.[22]

In principle, the (complex) observational biases in GRB detectors and spectrographic efficiency can be taken into account, but this is a substantial challenge. Since the total number of GRBs at low redshift is small (only a few below $z = 0.2$), we do not have an accurate accounting of the local rate of GRBs (beaming-corrected or otherwise). Obtaining a measurement of this rate requires an extrapolation from higher redshifts (where the rate is more accurately measured) and an assumption about how that rate is changing with cosmic time. Assuming that the overall long-duration GRB rate scales with the star-formation

Figure 4.3. (*Continued*) to improved detector sensitivity. (bottom) A model for observed E_{iso} in Swift-discovered GRBs accounting for intrinsic properties and observational biases. The curve shows the effective distribution of observed E_{iso} for different redshift ranges. The range $z = 1-3$ dominates the observed rate. Adapted from S. B. Cenko et al., *ApJ* **711**, 641 (2010); N. R. Butler, J. S. Bloom, and D. Poznanski, *ApJ* **711**, 495 (2010).

rate of the Universe,[*] the typical uncorrected rate density inferred around the Milky Way[23] is 0.1–1.5 GRB per year per Gpc^3. This means that the true rate (corrected for beaming) is about 5–500 per year per Gpc^3. This seems like a huge number[†] but it pales in comparison to the rate of other phenomena in the Universe. Supernova of Type Ib or Ic occur at about a rate of 20,000 per year per Gpc^3 meaning that the progenitors of long-duration GRBs, thought to be core-collapsed events like Ib/Ic SN progenitors (§5.1.2 and §5.1.1), occur at less than a few percent the relevant SN rate. Put another way, if core-collapsed stars are producing long-duration GRBs, such progenitors are a rarity even among such rare events: probably just about 0.1 percent of massive stars die producing long-duration GRBs. Though short-duration bursts are observed less often than long-duration bursts (§2.1.2), they are intrinsically fainter, so their true rates are thought to be somewhat higher (at $z \approx 0$) than long-duration GRBs. Interestingly, the rate of short-duration bursts is approximately comparable to the inferred rate of merging neutron stars (see §5.2).

[*] At low z, this assumption of scaling to the star-formation (SF) rate (SFR) is reasonable. However, at higher redshift, it appears that the GRB rate is enhanced relative to the star-formation rate, possibility due to metallicity effects. See §6.2 for a more detailed discussion.

[†] Especially given the fact that there is about $3,000 \, Gpc^3$ volume out to the redshift of GRB 090423.

5

THE PROGENITORS OF GAMMA-RAY BURSTS

> One day I undertook a tour through the country, and the diversity and
> beauties of nature I met with in this charming season, expelled every
> gloomy and vexatious thought.
> —Daniel Boone, from "Boone's Narrative,"
> appended to John Filson, *The Discovery, Settlement,
> and Present State of Kentucke*, 1784, 54–55

The physics of the central engine (§2.3) is constrained by basic measurements of GRB variability, inferred luminosity and energy release, and event spectra. This has led to the conclusion that a newly formed black hole or rapidly spinning neutron star lies at the heart of most distant GRBs: popular-level promotions of GRBs have noted, not unreasonably, that a GRB is the birth cry of a newborn black hole. But the tale we spin about GRB central engines may be more aptly seen as the denouement of an epic story about the tragic life cycle of the object(s) that produces the catastrophic event leading to the GRB. Studying GRB progenitors, then, is at once a quest to uncover the sorts of the preexplosion astrophysical entities that lead to a GRB and to understand how those entities evolved during their lifetime. Since clearly not all objects in the Universe make GRBs, we wish to understand what sets GRB progenitors apart from non-GRB progenitors.

Are the progenitors of GRBs, like Romeo and Juliet, fated to die so catastrophically from the outset, or are they just undistinguished actors who took some unfortunate turns in their life path?

By now it should be evident that there are few absolute certainties with GRBs; so it should come as little surprise that the understanding of the progenitors is no exception. One clear standout is the progenitors of Soft Gamma-ray Repeaters which are very obviously neutron stars. There are a number of corroborating lines of evidence for this progenitor association: (1) Some well-localized SGRs are associated with supernova remnants, suggesting they are byproducts of recent SNe (and many types of SNe are thought to leave behind NSs); (2) there is quiescent X-ray emission from the sites of SGRs, similar to a class of NSs called "anomalous X-ray pulsars"; (3) Galactic SGRs tend to be found in the Galactic plane (see figure 1.3), where most young NSs reside; and (4) the ringdown emission after SGR pulses is periodic, with periods comparable to that of slowly rotating NSs (few seconds). Moreover, there is good evidence that very high magnetic field strengths set the progenitors of SGRs apart from other neutron stars. The magnetic field strengths are inferred by observation of a slightly decreasing period between successive events, directly revealing the rate at which the progenitor appears to be spinning more slowly with time. One can calculate the amount of energy required to "brake" the progenitor's spin,[1] which, in turn, implicates magnetic fields of strength $> 10^{15}$ Gauss, tens of thousands of times higher than typical pulsar fields. The precise origin of the high-energy burst and the cause for the diversity of luminosities and light curves are not well understood.

5.1 A Massive-Star Origin

If magnetars are the most securely known progenitor of a subclass of GRBs, the next in line are massive stars, very likely the origin of some long-duration GRBs. We have seen (§4.2.1, §4.2.2) that the locations of long-duration GRBs around star-forming galaxies provide strong (but circumstantial) evidence for a massive-star origin of such events. Since the afterglows of long bursts appear to obliterate the incriminating evidence of their progenitor (§4.1.1), the best evidence comes in the aftermath of the GRB, as the afterglow is fading and the results of having blown up a star become manifest. We discuss this observational evidence in §5.1.2 but now turn our attention to physical models of massive-star progenitors.

5.1.1 Collapsars and Friends

The complexity of the life cycle of stars can be summarized as a long struggle between the attractive force of gravity on large scales and the repulsive forces of pressure on atomic scales. At high density and temperature, such as exists in the cores of stars, nuclear-fusion processes act as an abundant heat source, serving to keep the pressure high and halt the star from falling in on itself. For most of the life of stars, like a hot cold war, a happy and stable stalemate between the two forces persists. However, all stars eventually run out of nuclear fuel, causing the inner portions to contract. For stars less massive than about $8\,M_\odot$ at birth, the postnuclear burning phase contraction is halted by electron *degeneracy pressure*. An inert core of mass ~ 0.1–$1 M_\odot$—a white dwarf—remains essentially forever.[2] For stars more massive than about $8\,M_\odot$ at birth, the crush of gravity is so intense that electron degeneracy pressure is not enough to halt contraction. Instead, there is an implosion of the core, and, by various proposed mechanisms, this causes a massive shockwave to propagate outward through the collapsing star, causing it to explode. Densities and temperatures are so high in the regions of the exploding star that new elements, some radioactive, are quickly formed. Energy from the radioactive decay of some elements is deposited into the exploding material, and a supernova is born.

The specific type of supernova created in exploding massive stars depends both on the composition and size of the star before it explodes and the circumburst medium into which the outflowing material will propagate. The least massive "core-collapse" supernova progenitors have large hydrogen envelopes before explosion and leave behind an NS remnant. Higher-mass progenitors ($>25 M_\odot$ at birth) probably also leave behind an NS but, prior to explosion, have lost their hydrogen (a Type Ib SN) and, in the more massive cases, helium (a Type Ic SN). These hydrogen-stripped stars are called Wolf-Rayet stars, and there are many of them seen in the Milky Way. Understanding precisely how and why these stars explode is a continuing pursuit (outward-moving shocks from the creation of a newly formed NS and/or pressure from neutrinos are the main culprits). Also active is the art of modeling the diverse physical states in the supernova explosions themselves that give rise to the diversity of supernovae observed.

In 1993 Stan Woosley of the University of California, Santa Cruz, published a seminal paper highlighting the properties of a massive star ($15 M_\odot$ before exploding) that undergoes core collapse.[3] But instead of the creation of an NS at the center, he posited that a black hole could be formed under certain conditions. In this case, the usual mechanisms for exploding the star would not be strong enough to do so. This is why Woosley referred to this event as a "failed" Type Ib supernova. Mass from the star would instead flow inward to the newly formed black hole. If the star was rotating, then matter would fall freely along the rotation axis toward the black hole on timescales of seconds; however, along the equatorial plane, matter would be held up in an accretion disk by a centripetal force. This model, then, sets the stage for the central-engine scenario (§2.3) thought to power a long-duration GRB: a small region around a black hole where gravitational potential energy from an appreciable amount of mass ($> 0.1 M_\odot$) could be tapped over the course of seconds to tens of seconds.

Aside from the basics of the central engine, recall that relativistic outflow is a crucial requirement of the fireball (and afterglow) model. But too much matter entrained in the outflow—a so-called dirty fireball—would bog down the acceleration and stop the outflow from reaching relativistic speeds. To this end, it would seem that just about the worst place in the Universe to launch a pristine fireball would be at the center of a collapsing star. Indeed, the very crucial question of whether Woosley's model could also allow for relativistic outflow would not be answered for several more years. In 1999, Woosley and his student Andrew MacFadyen presented the results of a detailed numerical simulation that followed the collapse to a black hole, the dynamics of the accretion disk, and the fate of the energy deposited in the region around the polar axis.[4] They, and subsequently others, showed that the polar regions* become sufficiently evacuated of matter (after a few seconds) to allow the deposited energy to begin to propagate outward into those regions. The in-falling matter outside the polar regions creates an effective barrier that funnels the outflowing material into a jet. This jet, which will last as long as matter continues to feed the central black hole, in turn pushes aside any remaining matter. By the time the jets emerge, simulations have shown that they can reach the requisite Lorentz factors ($\Gamma \sim 100$; see footnote on page 56) and that they are highly collimated (opening angles of a few degrees, as inferred from afterglow modeling; see §3.2).

The launching of the jet has consequences for the star not entirely anticipated in the original 1993 picture. First, the jet will interact with the matter in the jet boundary,

---

*The polar axis is the direction that is perpendicular to the rotation of the accretion disk, probably similar to the overall polar axis of the progenitor precollapse. The two polar regions are the areas around the polar axis both above and below the central engine (think of two ice-cream cones attached at their tips at the central engine).

creating shocks within the outflow. Therefore, even if the energy input is constant near the black hole (which it should not be, in general), these shocks lead to an unsteady relativistic "wind," also a basic requirement of the GRB production mechanism in the internal-shock scenario. Second, the jet is thought to carry enough energy that, even if a small fraction of it couples to the matter in the jet boundary, it will be enough to explode the star. During this disruption, explosive *nucleosynthesis* (particularly of the element ^{56}Ni) can occur in the dense regions of the jet boundaries and also (more likely) from a wind produced in the outskirts of the accretion disk. Explosion of a star and nucleosynthesis of ^{56}Ni are the ingredients of a supernova, so, contrary to the original "failed Ib supernova" scenario, it became clear on theoretical grounds that the more sophisticated collapsar model indeed predicted both a GRB (from the polar jets) *and* a supernova.

5.1.2 Connection to Supernovae

The particular flavor of supernova that should accompany a long-duration GRB in the collapsar progenitor model is somewhat of a postdiction: by 1999, GRB 980425 had already been phenomenologically connected to a strange supernova SN 1998bw. That supernova was very bright, comparable to some of the brightest supernovae ever observed at the time. This indicated a large amount of ^{56}Ni synthesis: at least 0.5 $M_\odot$ in just that element alone. The SN spectrum also showed broad undulating features indicative of high-velocity outflow (owing to large Doppler shifts in the ejecta), at least several tens of thousands of kilometers per second (an order of magnitude faster than most other SNe). Importantly, the spectrum showed no strong evidence for the presence of hydrogen or helium in the outflowing material. These observations led to the observational classification of SN 1998bw as a broad-lined Type Ic (Ic-BL) supernova. The radio light curve was unlike that seen in any SN to date and was modeled as having arisen from mildly relativistic outflow in the outer layers of the ejecta.[5]

Theoretical models explaining individual SNe derive from an admixture of basic physical equations, intuition, and computationally intensive supercomputer modeling of the complex interplay between light, gas, and various heating sources. For SN 1998bw, a model[6] developed to explain the brightness, apparent velocity, elemental abundances, and time evolution suggested that a hydrogen- and helium-stripped star (a so-called carbon-oxygen star) of mass $\sim 14\ M_\odot$ produced a whopping 0.7 $M_\odot$ of ^{56}Ni; the star would have been $\sim 40\ M_\odot$ at birth. Since the total energy coupled to the ejecta was $E_{SN} \approx 10^{52}$ ergs (roughly a factor of ten larger than the typical energies of SNe), SN 1998bw was advanced as a new class of exploding star called "hypernovae," connoting especially energetic supernovae.

The progenitor of SN 1998bw appeared to be very similar to what Woosley had envisioned in 1993. But the substantive difference in the outcome (aside from the fact that an SN did indeed occur) was that a very weak GRB was produced. Indeed, GRB 980425 was at least three orders of magnitude less energetic than the "typical" cosmological long-duration GRB with known redshift. The extreme nature of both the GRB and SN threw many for a loop, but the collapsar proponents quickly saw this odd couple as possible bookends of a more typical population. In this case, much of the energy reservoir available during core collapse was coupled to the protons and neutrons in the star, giving rise to the observed fast-moving (but mostly nonrelativistic) ejecta in the SN; a minority fraction of the energy was coupled to the mildly relativistic material (as manifested by the radio light curve) and the highly relativistic material (as manifested by the GRB). But in other cases, presumably with cosmological long-duration GRBs, there could be a different partitioning of the energy. By 1999, Woosley and MacFadyen suggested that all long-duration GRBs should be accompanied by supernovae with qualities similar to SN 1998bw.

The search was on for SNe associated with cosmological GRBs. I presented evidence[7] for an SN-like bump in the light curve following the event GRB 980326. While the timescales were similar to SN 1998bw and the colors appeared consistent, no easily recognizable signature of an SN was seen in the low-quality spectrum taken around the time of maximum light of the bump. Likewise, since no redshift was known for the GRB or host galaxy, the evidence for an SN was far from ironclad. Shortly after publishing our results, a credible SN-like bump was found in the reanalyzed afterglow light curves of GRB 970228. Over the next several years, evidence of bumps was found buried in many GRB afterglows, with events like GRB 011121 showing strong photometric evidence for a 1998bw-like supernova. Some events appeared even brighter than SN 1998bw and some fainter (by as much as an order of magnitude in peak flux).

The smoking gun that would convince most remaining skeptics of the connection between GRBs and the death of massive stars came in 2003. GRB 030329 was an incredibly bright event with a bright afterglow, originating from one of the lowest redshifts to date ($z = 0.16$). By most metrics, it was an ordinary long burst of duration $T_{90} \approx 23$ seconds.[8] Yet, as the afterglow faded, a set of broad spectroscopic features—subtle at first, then overwhelming—developed. Removing a model for an afterglow component from spectra taken eight days after the event, Tom Matheson and Krzysztof Stanek (both then at Harvard) and collaborators found striking evidence that these features resembled the high-velocity features of SN 1998bw (figure 5.1). Other groups confirmed the remarkable finding: another broad-lined Type Ic supernova was associated with a GRB, but this time it resembled the other cosmological events that dominate the observed rate of GRBs.[9]

By July 2010, three additional events (from low redshift, when the SN would be expected to be detectable) showed strong spectroscopic evidence for an SN component, and many more had strong photometric evidence for such. However, most Swift GRBs do not have associated SNe, and the explanation for this is straightforward: even if SNe accompany all long-duration GRBs, these types of SNe appear exceedingly faint at optical wavebands when they occur beyond redshift of $z \approx 1$. This is due to the suppression of the blue and ultraviolet portions of their

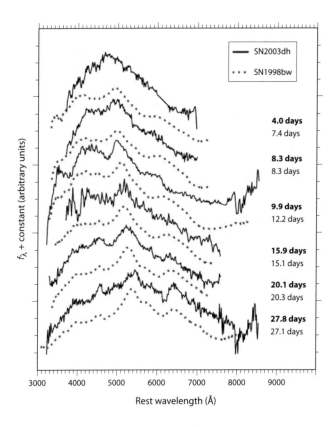

Figure 5.1. Comparison of the spectral evolution SN 1998bw and SN 2003dh. Solid lines show SN 2003dh, associated with GRB 030329, and dotted lines show SN 1998bw taken at similar times since the GRB trigger. Times after the GRB are given in the restframe, correcting for cosmological-time dilation. The remarkable similarity between the two supernovae showed that at least some cosmological GRBs arise from the same type of progenitor as the oddball GRB 980425. Adapted from J. Hjorth et al., *Nature* **423**, 847 (2003).

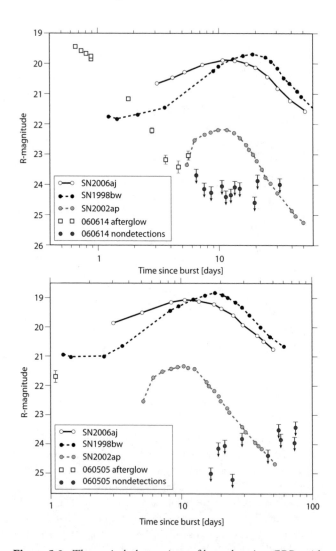

Figure 5.2. The optical observations of long-duration GRBs without associated SNe to deep levels. The gray points are the detected points in the afterglow light curve; the points with a

(*Continued*)

spectra (redshifted into the optical bandpass for $z \gtrsim 1$) due to heavy metals (e.g., iron) in the ejecta. Moreover, since GRB-SNe are about as bright as the galaxies they occur in, we contend with both afterglow light and host galaxy light in searching for the SNe.

5.1.3 Long-Duration Events without Supernovae

This notion that all long-duration bursts *could* be associated with Type Ic supernovae persisted for years. But in the summer of 2006, two GRBs were discovered by Swift at sufficiently low redshift that the SNe, if they were of similar brightness to SN 1998bw, should have been easily detected. Instead, despite concerted efforts to find the SN peaks in GRB 060505 and GRB 060614, no such signatures were found in deep images (see figure 5.2). Using some prior inference on the lack of dust obscuration from afterglow fitting, a Danish-led collaboration (which included me) suggested that any contemporaneous SN must have been fainter at its peak brightness than about -14 magnitude; this magnitude translates into a

---

Figure 5.2. (*Continued*) downward-facing arrow are nondetections ("upper limits") at the position of the afterglow, meaning that any source at that position would need to be fainter. Overplotted are three GRBs as they would have appeared at the redshift of the GRB (SN 2006aj, associated with GRB 060218; SN 1998bw, associated with GRB 980425; SN 2002ap, a faint broad-lined Ic SN). All SNe like these fiducial events are ruled out by the data. Adapted from J. P. U. Fynbo et al., *Nature* **444**, 1047 (2006).

luminosity less than about one-tenth to one-hundredth the peak luminosity typical of most SNe, essentially ruling out all known types of SNe.[10] Some suggested that the events were simply "short-burst impostors" and, therefore, on physical grounds that we should not have expected an SN (see §5.2.2). Others suggested that the GRB might have originated from much higher redshift than the putative host galaxies, appealing to a chance coincidence of the alignment of those distant GRBs with foreground galaxies.[11]

My take, which is by no means universally accepted, is that instead of invoking the need for "long-duration short bursts" (syntactically discordant, at the very minimum) or multiple chance superpositions with foreground galaxies, the natural explanation for such events[12] would be that they did indeed arise from the collapse of massive stars, but instead of producing the necessary conditions to blow up the star and produce a supernova, only a GRB was produced. In a sense, I suggest that Woosley's original "failed supernova" hypothesis may have been realized* with these events.

---

*When Albert Einstein looked at cosmological solutions to his General Relativistic field equations, the solutions he found told him that the Universe was either expanding or contracting, something that contradicted the observations of the day. So he added in a "cosmological constant" to create a steady universe. Soon after it was shown that the Universe was indeed expanding, Einstein is famously quoted as saying that the addition of that constant was the "biggest blunder" in his professional career. Six decades later, evidence for a nonzero cosmological constant was uncovered by detailed measurements of SNe (see §6.5). This essentially vindicated Einstein's tweaked equation and added deep irony to his "biggest blunder" exclamation. With affinity and affection, I think it is apt to call this interpretation of supernova-less long-duration GRBs as "Woosley's biggest blunder."

All theoretical studies of collapsars have noted the importance of the angular momentum of the precollapse star. With too little momentum, an accretion disk feeding the central black hole cannot be supported long enough to power a jet through the collapsing star. Spinning too fast, the accretion disk is insufficient to feed the black hole and power the GRB. Since the total angular momentum before collapse is a complex function of many competing physical mechanisms, it is essentially a free parameter in the prescription for events that follow core collapse. One possible explanation for the origin of GRB 060505 and GRB 060614, and more generally in the diversity of GRB-SN brightnesses and light curves, is simply that the differences in the angular momentum affect the total ^{56}Ni production.

To be sure, rotation of long-GRB progenitors has yet definitively to be established observationally as a driver of GRB-SN diversity. Indeed, on the theoretical front, the amount of rotation required for the nominal collapsar model to produce a GRB and an SN may be difficult to arrange in practice. In envisioning the full life cycle of the massive star that leads to the pre-core-collapse progenitor, we require that a significant amount of mass be lost from birth to death (perhaps as much as 25–$40\,M_\odot$). Most stellar mass loss happens in stellar winds, and Wolf-Rayet stars (in particular) lose copious amounts of mass, at a rate of about $1\,M_\odot$ every ten thousand years! But mass ejected to large distances is very effective at slowing up what remains of the star.[13] Therefore, "normal" channels of mass loss for massive stars might lead to evolved stars that are spinning too slowly and thus are incapable of making GRBs. To this end, some have advocated that a productive mass-loss channel involves the interaction of the progenitor with a binary companion star. In this case mass can be flung from the progenitor, but the angular momentum can remain relatively high (where the spin rate would now be dictated by the orbital period of the binary system). How angular momentum is maintained in GRB progenitors is an open question. It is also an open question how important binary-star interactions are for GRB/SN production.

5.2 Mergers of Compact Objects

Historically, there were many attractive features of merging compact stars as GRB progenitors.[14] First, we have known for decades that such systems exist in the Milky Way and that, in the case of some observed NS–NS binaries, the merger will happen in less than the current age of the Universe. Second, the modeled (albeit uncertain) rates of coalescence of such systems approximately match the inferred rate of GRBs. Third, the variability timescales in GRBs are naturally explained by the light crossing time of the two objects at the time of merger. Fourth, the total mass-energy available to be accreted into a newly formed black hole is larger than that required to produce the total energy released in the GRB event itself plus the afterglow. Indeed, merging NSs were the leading explanation for cosmological GRBs until a pernicious Mother Nature decided to reveal long-duration GRBs as arising from an entirely different beast.

By 2004, with the localization of short bursts missing from the observational trophy chest, degenerate merger events remained a viable progenitor scenario. Indeed, one of the "hopes" of Swift—vindicated soon after launch by observation—would be that regular Swift localizations of short bursts would help reveal their nature. Though we are not nearly as certain as the progenitors of SGRs or long-duration bursts, observational evidence is mounting that this subclass may indeed be due to such progenitors.

5.2.1 Models

The stellar graveyard consists of just a few classes of remains or compact "remnants": white dwarfs, neutron stars, and black holes.[15]* Getting to the central engine configurations in models involving such objects is tantamount to envisioning a scenario where these objects play primary roles both in rapidly supplying the necessary mass and in serving to channel that mass into the energy that powers a GRB (see figure 1.4 for an illustration). Since the pathways to creation of these remnants are numerous, the degenerate progenitor models must also chart the

---

*If these end states of stars are arm-wrestling matches between gravity and opposing forces: white dwarfs (supported by "electron-degeneracy pressure"; §5.1.1) and neutron stars (in part supported by "neutron-degeneracy pressure") represent an eternal stalemate of sorts. But with black holes, gravity has gone over the top to pin its opponents: all pressure support is insufficient to counteract the crush of gravity; therefore, all the mass of a black hole is concentrated at one point. It is thought that quantum mechanics plays a role in halting the concentration of all mass to an infinitesimally small region of space. So the mass in a BH may be concentrated in a finite volume.

long-term evolution of the objects that ultimately lead to the central engine.

The progenitor scenario most discussed in the literature involves the coalescence or merger of two neutron stars. One channel, probably the dominant mechanism, for the creation of this merger event starts the story with a binary system of two newly formed high-mass stars. Unlike the low rate of human twin birth, it is very common for stars to be formed in binary systems. Since stars born in binaries tend to have similar masses, this also implies that the stars will undergo a similar arc of development. If one star undergoes a core-collapse SN, it will leave behind a neutron star. We believe that NSs born in SN explosions undergo a "kick" during their creation that sends them flying off in some random direction relative to where the presupernova star was heading. In some cases, the NS will have a large-enough velocity simply to bid its sibling *adieu* and disrupt the binary. But in many cases, the velocity of the NS will not exceed the critical velocity needed to escape the gravitational pull of the system. In this case, the binary system will persist. What happens then depends on the state of the other star and the details of the binary orbit.[16] In some cases, the companion star may also explode as a core-collapse supernova, leaving behind an NS. In most cases, the binary system will disrupt but with the conditions of the kick just right, the NS–NS binary will survive.

At this point, the neutron stars orbit around each other in a highly stable configuration, interacting only through their mutual gravitational attraction. Isaac Newton would have posited that these two objects could dance the

"Do-Si-Do" for eternity. But General Relativity introduces a certain fatigue to the system: over time, the binary orbit begins to decay as energy is radiated through gravitational waves. Eventually the binary decays to the point where the NSs collide and a GRB central engine is born. Much of their mass goes into making a spinning black hole with mass 2–3 $M_\odot$. Some of the mass forms an accretion disk (lasting < 1 sec) that feeds the black hole, and some of the mass (0.1–0.2 $M_\odot$) is flung out.

How often this NS–NS merger scenario occurs is not precisely known; but, given that we know of ∼6 NS–NS systems in the Milky Way that appear to have formed this way, it seems rather clear that it *does* happen. Since we also have observational evidence of the binary fraction of massive stars and of the distribution of kick velocities that NSs receive on birth (during supernovae), attempts have been made to predict the rate of NS–NS mergers using such observational ingredients. Current estimates place this rate (in the Milky Way) to be between one event per million years to one event per hundred thousand years.[17] Other "channels" for NS–NS production may be important in the overall rate of coalescence. For instance, in some sort of cosmic square dance, neutron stars populating dense regions of stars may actually exchange binary partners. Globular clusters—conglomerations of millions of stars and thousands of neutron stars—serve as the main dance floor for NS–NS binaries formed this way.

There is a rich literature describing other viable compact objects as GRB progenitors. The double supernova channel, for example, may also produce neutron star–black hole (NS–BH) binaries that will also coalescence after orbital decay due to gravitational-wave emission. In this case, one of the two initial stars may be massive enough to produce a BH directly when it explodes as a supernova.[18] Or the first neutron star could accrete enough matter from the second star to collapse to a BH before the second star explodes as a supernova.

5.2.2 A priori expectations

Since the viable compact merger scenarios essentially lead to the same (short-lived) central engine, there is no obvious observational path to distinguishing between the progenitor possibilities from the GRBs or afterglows. However, there are several sets of observations that should winnow down the choices among the degenerate-binary merger scenarios. Most striking are the different a priori expectations of degenerate-binary merger scenarios relative to the expectations from collapsars. First and foremost, none of the compact progenitor scenarios should lead to a bright Type Ic SN, so the presence of a contemporaneous SN would rule out such scenarios for that GRB.* We can view the absence of an SN as a necessary but not sufficient condition for confirming such progenitors.

Host Galaxies: For some of the possible progenitors, particularly NS–NS mergers, the broad distribution of merger times since birth implies by the time of the GRB, that the general population of stars in the host galaxy will

---

*Of course, as the discussion in §5.1.3 asserts, the absence of an SN does not rule out massive stars, nor does it require degenerate-binary merger progenitors.

have evolved appreciably. In particular, since star formation appears to happen in "bursts" of time lasting tens-
of-million to hundreds-of-million years, the star-forming
episode that gave rise to the progenitors will have subsided,
and the stars in the host will have aged significantly. Thus,
GRBs from NS–NS mergers are expected to be associated
with more of a diverse population of host galaxies than
those hosting collapsars. This diversity would obviously
be reflected in average stellar age and, most manifestly, in
galaxy color.

Locations: Since NS–NS progenitors also experience
a systemic velocity "kick" during formation of upward of
several hundred kilometers per second ($v_{kick} \approx 300$ km/s),
we might expect them to travel far from their birth site
in the time τ ($\sim 10^{8-9}$ yr) until merger. This is quite
different from collapsars that should not travel far before
core collapse. In linear distance, this is

$$l \approx v_{kick} \times \tau = 30 \text{ to } 300 \text{ kpc.} \qquad (5.1)$$

This distance traveled can be much larger than the size of
the host galaxy; in some cases v_{kick} may be larger than the
velocity required to escape the gravitational tug of the host
galaxy.[19] Therefore, we naturally expect NS–NS binaries
to be loosely distributed around their host galaxies and,
in some fraction of the cases, may be far from the light
of their true host. NS–NS binaries literally evaporate from
galaxies. Other variants, like NS–BH binaries, may receive
less of a kick during formation and thus, like high-school
sweethearts who get married quickly and never leave their
hometown, will die near to their birth site.[20] Likewise,

short-lived progenitors, even if they travel fast, might not make it far from their birth site until the coalescence. Given these considerations, we expect that some GRBs from compact binary progenitors should not be significantly offset from the light of their host galaxy. However, since we do not know from *ab initio* models precisely what the merger time nor the kick distributions will be, we cannot predict precisely what the "offset distribution" of events from such progenitors should be. But compared to collapsars, most models do generally predict a wider separation of such progenitors from galaxy hosts.*

Short-lived Transients: If some events do indeed occur far from the stellar birth site, they might also occur far from the regions where there is an appreciable density of ambient gas. Under the assumption that external shocks dominate the emission of afterglows, this implies that such events should have a relatively weak afterglow. Indeed, in the simplest models, for a fixed energy in the external shock, the peak brightness of the afterglow at a fixed wavelength should scale as the square root of the ambient density. Given that the density in the intergalactic medium can be millions of times less than the density in the interstellar medium of galaxies, we might expect that afterglows from such progenitors could be upward of a thousand times fainter than afterglows from events produced within galaxies.[21]

---

*Still, locations alone certainly cannot be the ultimate indicator of collapsar or compact merger progenitors: we know, for instance, that core-collapse supernovae can happen in small satellite regions of star formation far from galaxies. Likewise, some collapsars might be found far from the detectable light of distant galaxies.

In most merger models, the mass that does not flow into the newly formed black hole may instead get ejected from the system. In the case of a shredded NS, we expect the flung-off mass to undergo a rapid burst of nucleosynthesis, given that it is highly compressed and rich in neutrons. The decay of these rapidly produced radioactive isotopes of heavy elements serves as a powerful heating source for the outflowing material. Just as in a supernova, the heat generated by radioactive decay is trapped by the dense ejecta until the material expands to a much larger radius. It takes a few weeks until an ordinary supernova reaches maximum brightness, but these events, given that there is much less mass involved, will reach peak brightness about one day after the coalescence. The nominal expectations for such a short-lived radioactive-powered transient was worked out by Bohdan Paczyński and Li-Xin Li[22] in 1998. More detailed modeling of the nucleosynthetic processes that drive these "Li-Paczyński mini-supernovae" has been carried out recently in much more detail.[23] The conclusion is that there should indeed be a short event reaching peak visual magnitude of −14, about one hundred times fainter than SN 1998bw. Saying "Li-Paczyński mini-supernovae" is a bit of a mouthful (say it ten times, fast!), so several alternative names have been proffered. Since the peak of the event is expected to be approximately one thousand times brighter than a nova event, I prefer the term "kilonova," advanced by my colleague Eliot Quataert at the University of California, Berkeley.

Gravitational Waves: The last on our list of a priori expectations from compact mergers is gravitational waves

(GWs). Just as accelerating charged particles produce light, GWs are produced by accelerating mass (in certain geometric configurations). GWs are ripples that deform space and time and travel at the speed of light. The slow procession to coalescence of widely separated compact binaries occurs because gravitational waves gradually carry away energy from the system. But in the last second before coalescence, the rate of acceleration increases dramatically, and a tremendous burst of gravitational waves emerges, hastening the coalescence. For a distant observer, the GRB would be then be accompanied by a (near) simultaneous gravitational-wave event. Given the masses of the coalescing objects and the distance to the event, the intensity and temporal development of the gravitational-wave signal can be calculated reasonably precisely. As GRB-SNe were for collapsars, the discovery of a GRB and concurrent gravitational-wave event is considered the observational smoking gun that would unequivocally reveal the precise nature of the progenitor of that event (i.e., a compact binary merger). We discuss this possibility further in §6.5.

5.2.3 Confrontation with Observation

As a group, observations of long-duration GRBs do not appear to reflect these a priori expectations of compact-binary merger progenitors. But, as shown throughout our discussion of the environments of GRBs (chapter 4), the types of galaxies that are associated with short-duration GRBs and the locations of the events around those galaxies appear to reflect some of the compact-binary merger-model expectations. For example, whereas no long-duration GRBs appear to be associated with galaxies dominated by old stars, 10–20 percent of short-duration GRBs are associated with old-stellar-population galaxies (such as ellipticals). Similarly, as shown in figure 5.3, the locations of short-duration bursts appear to be more widely distributed around their associated hosts. However, unlike with long-duration GRBs, the association of a given short burst with a given host galaxy tends to be more uncertain. The reason for this is partly due to the observational bias of poor afterglow localizations[24] and partly, it seems, intrinsic. Indeed, viewing the question from the model perspective, if we posit that some fraction of merger events should happen far from their host galaxies (§5.2.2), we *expect* that a fraction of events should have ambiguous (or incorrect) host galaxy associations and thus incorrect offset measurements. This is like letting a bunch of cats loose in a crowd and, after some time has passed, saying that the owner of each cat is the person it happens to be nearest to.[25] The short GRB 060502b is emblematic of the difficulty of host association.[26] The event was 90 milliseconds (0.09 sec) in duration, had an X-ray localization of 4.4 arcseconds, and had no detected optical afterglow. In the error region there are a few faint (presumably distant) galaxies that could be the host. But 17 arcseconds away (i.e., relatively nearby) there is a bright-red galaxy made up of lots of old stars. At the distance of that galaxy, the angular separation on the sky corresponds to a physical separation of 73 ± 19 kpc in projection. This is a distance easily obtainable for NS–NS binary coalescence (equation 5.1) and comparable to that

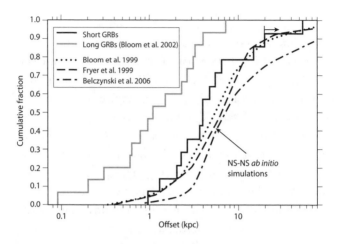

Figure 5.3. The cumulative location distribution of short- and long-duration GRBs around their putative host galaxies. The twenty (pre-Swift) long-duration GRBs (light gray line; from J. S. Bloom, S. R. Kulkarni, and S. G. Djorgovski, *AJ* **123**, 1111 [2000]) all lie within about 10 kpc from the apparent centers of their hosts. In contrast, short-duration bursts (even selecting those with the most clear associations to individual galaxies) appear to reside at larger distances from their hosts. Depicted in smooth curves are three different *ab initio* models for the location distribution of merging NS–NS binaries, showing reasonable agreement with the observed short-burst population. Adapted from W. Fong, E. Berger, and D. B. Fox, *ApJ* **708**, 9 (2010).

inferred for GRB 050509b from a very similar elliptical galaxy. Without a redshift of the event (which could have easily disproven the elliptical galaxy association), we are left with an almost unpalatable uncertainty about the host and location from it. Everything we assert in astrophysics is probabilistic in nature, but connecting short-duration bursts to galaxies is a particularly uncomfortable exercise.

To date no short-duration bursts have had detected SN emission, but only a handful of short bursts (with reasonably secure redshift) have had deep-enough follow-up to rule out SNe beyond a reasonable doubt. Of course, the absence of SNe in these events is certainly tantalizing but by no means definitive. Likewise, kilonovae—one rather clear predicted signature of compact mergers—have not been definitively identified. There was a curious short-duration GRB (080503) that showed evidence for a one-day timescale optical event (see figure 5.4) that nominally fit kilonova models.[27] However, since no redshift of the source could be established, the intrinsic peak brightness could not be calculated. More importantly, there was detected X-ray emission that appeared to track the optical event at around one day after the GRB, something not predicted in kilonova models. Our conclusion from this event was that a kilonova was indeed a possible explanation but not the only viable model. Taking the short-burst population as a whole, no other searches reported to date have been sensitive enough to detect such a faint event, so it is difficult to draw any strong conclusions. Future searches for kilonovae (both triggered by GRBs and perhaps found in untriggered optical surveys) should prove most illuminating.

A concurrent gravitational-wave event—the cleanest and most distinct prediction from compact mergers—has not be confirmed observationally. But since current gravitational-wave detectors are not sensitive enough to NS–NS events beyond the distance of the very nearest galaxies (such as Andromeda), the rate of NS–NS (or NS–BH) mergers is thought to be simply too low to

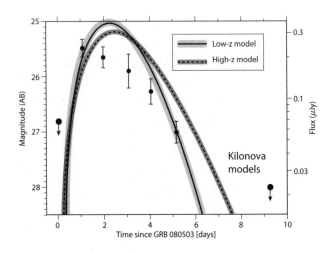

Figure 5.4. Kilonova model fits to the curious late-time emission from short-duration GRB 080503. The characteristic timescale of one day for the time to peak is related to the velocity of the outflow and the mass of the ejected material from an NS–NS merger. The luminosity/magnitude is related to the total mass of the ejected material and the radioactive heating rate (a quantity difficult to determine theoretically). Unfortunately, the redshift of GRB 080503 could not be inferred from nearby associations, so two models are plotted ("low z": $d = 124$ Mpc; "high z": $d = 2.7$ Gpc). The basic characteristics of the expected kilonova light curve are well matched, although in detail there are discrepancies (not, at this point, all that worrying given the large uncertainties in the models). Adapted from D. A. Perley et al., *ApJ* **696**, 1871 (2009).

detect even one event per decade with the current setups. The good news is that more sensitive searches are on the horizon (see §6.5) and that we may indeed detect definitive gravitational-wave signatures of short-duration GRB progenitors as early as ~2015.

The evidence, albeit circumstantial, continues to mount that at least some short-duration GRBs have *different* progenitors than do long-duration GRBs, and it seems clear that these progenitors are associated with an older stellar population than collapsars are. There is no unambiguous evidence to date to implicate NS–NS or NS–BH binaries as short-duration GRB progenitors, but the general consensus by 2010 was that this class of progenitors seem to be the most likely.[28]

While kilonovae and afterglows are a natural consequence of the properties of the outflow, in the NS–NS or NS–BH scenarios, all known timescales for the mass flow in the merger configuration are very short: there are no natural timescales for the inflowing matter of more than a few seconds. The a priori expectations, then, are that we should see no significant emission on tens-to-hundreds of second timescales and beyond. Yet long tails of emission extending to hundreds of seconds after the GRB are often observed in what otherwise appear (in the initial spike) to be ordinary short bursts.[29] Moreover, the late-time X-ray flare in the short GRB 050709 (§3.1.1) finds no natural explanation in the NS–NS or NS–BH models. To be sure, many short bursts show neither long-lived tails following the prompt emission nor X-ray flares. On the other hand, those short bursts that do exhibit such behavior remind us to keep an open mind about the progenitors of all short-duration GRBs.

5.3 Extragalactic Magnetars

The bright flare from SGR 0525−66 during the March 5th event was outclassed by even brighter flares from other

SGRs. The brightest known giant flare from an SGR—an event from SGR 1806−20—occurred on December 27, 2004. It had a total energy release of few × 10^{46} erg, a factor of more than one hundred brighter than the March 5th event (but still a factor of one hundred less energetic than the underluminous GRB 980425). At the luminosity of the very brightest giant flares from magnetars, we should be able to see such flares from nearby galaxies. But at such large distances we would only see the initial hard-spectrum spike of emission and miss, below the detection threshold, the characteristic periodic ringdown seen in more nearby Galactic SGRs flare events. That is, giant flares from SGRs in other galaxies would appear like a lot like short-hard GRBs.

Even with a few nearby examples, the typical energies and the volumetric rate of giant flares from SGRs are difficult to pin down precisely. The events are rare (probably no more than once per 5000 years per source), and the distances to most SGRs is uncertain (leading to an uncertainty in the intrinsic brightness of detected events). By correlating the BATSE and IPN samples with nearby galaxies, it appears that a few percent (and no more than 10 percent) of the short events observed with BATSE may have occurred from extragalactic magnetars. No Swift event has been convincingly associated with a "local universe" galaxy (<200 Mpc)[30] but, during the Swift era, two IPN short-hard bursts were localized close to very nearby neighbors of the Milky Way: GRB 051103 (near galaxy M81 at a distance of 3.5 Mpc) and GRB 070201A (near Andromeda, at a distance of 770 kpc). Neither of these events showed any evidence for periodicity in the light curves. On statistical grounds, it is likely that at least one of these events did indeed originate from its putative host.

Accepting that at least one event originated from nearby, what was the progenitor? Without a supernova detected to very deep limits, we can certainly say they were *not* due to an ordinary collapsar. By great fortune, the *Laser Interferometer Gravitational-wave Observatory* (LIGO) project was collecting science data at the time of GRB 070201A. The LIGO team was able to put very useful limits on the lack of gravitational waves from the direction of Andromeda around the time of the GRB. The LIGO data were enough to rule out NS–NS and NS–BH mergers to a high degree of confidence up to the distance of Andromeda. This leaves extragalactic magnetars as the most likely suspect. The implied energy release at the distance of Andromeda was indeed close to that inferred from the December 27th event from SGR 1806−20, providing some confirming evidence. However, there was no obvious supernova remnant in the IPN error region of GRB 070201A. This suggests either that the NS traveled far from its birth remnant or that the magnetar was produced by other means.[31]

5.4 Classification Challenges

Nature has clearly found numerous ways to express herself on the canvas of high-energy photons.* It is, in many

---

*This is even more true if we do not restrict ourselves to events that look similar to GRBs (i.e., those events populating the traditional $T_{90}-E_{peak}$ plane). Among transient events, we know that terrestrial lightning produces bursts of gamma rays

respects, our calling to try to uncover the physics and origins of these different progenitors and the resulting burst mechanisms. But to understand the various progenitors in individual classes, we presuppose that we can indeed classify events *physically*. Instead, all we know for sure is what we observe—all we know we can do is classify *phenomenologically*. Physical classification is not just the ultimate goal: it has immediate consequences on the scientific process itself. In particular, resources for following up the bevy of well-localized GRBs are scarce, and it is clear that most of the juiciest science comes only when follow-up is robust. Given some foreknowledge of the likely physical classification (i.e., "this event might be very low redshift and thus allow us to observe the GRB-SN in great detail" or "this event might be very high redshift"), different researchers will respond with their resource allocations in different ways.

At first blush, it would seem that the two loci in the hardness–duration space provide a direct observational conduit to two distinct physical classes: long-soft bursts are collapsars, and short-hard bursts are due to, for instance, NS–NS mergers. Indeed, in recent years, calling an event a "short burst" (even if it has a long-duration tail of high-energy emission!) has become synonymous with "it originates from a compact binary merger" (and vice versa); likewise, calling an event a "long burst" is shorthand for saying it "originated in the core collapse of a massive star" (and vice versa). But in the presence of extragalactic

as do events near the surface of the Sun. Some massive black holes at the centers of galaxies produce episodes of high-energy gamma-ray emission, and so forth.

magnetars "polluting" the short-hard burst locus, we see a clear example of where this one-to-one mapping between what is observed and what is inferred breaks down. In the long-soft burst locus, we also have possible evidence (in GRB 060505 and GRB 060614; §5.1.3) of pollution by a noncollapsar population. Other long-duration events also hint at a greater physical diversity. For instance, GRB 070610 ($T_{90} = 8.3$ sec) was localized near the Galactic plane, and its afterglow flickered strangely for days. We do not yet know what the progenitor was, but it almost certainly originated from within the Milky Way and was unlikely to be a collapsar.[32] Even if a given GRB could be somehow classified intrinsically by its high-energy emission alone, the range of redshifts at which GRBs originate induces some important observational blurring. For example, a short-hard burst originating from high redshift might appear long and soft. Likewise, a long-duration GRB with an initial spike of emission might only have the initial spike detected above the instrumental threshold, leading to a classification as a short-duration GRB.*

Some have introduced new metrics on GRB light curves as a way to disambiguate between classes, such as focusing on the temporal and spectral properties of the initial pulse of the event. But clearly a GRB is so much more than the initial deluge of gamma-ray light: it is the afterglow, it is the supernova that may or may not follow, it is the emission in gravitational waves and neutrinos. Indeed, we

---

*Of course, comparing events using restframe quantities relies on a redshift measurement, which is not always possible.

have seen that the total energy release in the GRB itself ($E_\gamma \approx 10^{48-51}$ erg) is, in the case of long-duration bursts, probably less than the energy associated with the GRB-SN ($E_{SN} \sim 10^{52}$ erg). Moreover, in many progenitor models, this release is almost certainly less than the energy radiated in the form of neutrinos and gravitational waves (which have not yet been detected). So trying to devise a physical classification for a GRB from gamma rays alone might be like classifying the character of a person by studying her nose.

We have been looking at a deep intermingling between what is observed and what is inferred about underlying causes. This healthy conflation lies at the heart of many other fields in astronomy (supernova and quasar classification spring to mind). Indeed, the challenges of classification, especially in the face of so much diversity, might be taken as an essential component of astrophysics, a manifestation of the inexorable connection between observation and theory.

6

GAMMA-RAY BURSTS AS PROBES
OF THE UNIVERSE

The cosmos is full beyond measure of elegant truths; of exquisite
interrelationships; of the awesome machinery of nature.

—Carl Sagan, "The Shores of the Cosmic Ocean"
episode of *Cosmos: A Personal Voyage* (1990)

Until now we have concentrated on the observations and
theory of GRBs. The events have shown themselves as a
complex, panchromatic phenomenon that can arise from
a variety of different progenitors in both the nearby and
distant Universe. The GRB community has succeeded
in measuring the diversity of basic parameters of the
explosions and clearly learned a great deal about the
physics of the emission mechanisms. But despite a raft
of certainties, we could fill a book with what we do not
know. Indeed, filling in these sometimes gaping holes and
searching for new insights into the phenomenon keep
many of us awake a night.*

In parallel with the ongoing study of these events
and progenitors, new lines of inquiry have burgeoned:

---

*This is both figuratively and literally true: GRB practitioners on the front lines
of observation are often awoken at odd hours of the day and night by text
messages from satellites announcing new GRB discoveries. GRBs are terrible
strains on interpersonal relationships!

using GRBs as unique probes of the Universe in ways that are almost completely divorced from the nature of GRBs themselves. It may, at first, seem born of an ill-conceived crusade to admit that GRBs might tell us something interesting and deep about the Universe if we are willing to admit a large measure of consternation about our understanding of the probe itself. But the history of astronomy is littered with similar (apparently quixotic) endeavors. Pulsars are a prime example: we know that they come from spinning, magnetized neutron stars, but the precise emission physics responsible for the complex phenomenology is not well understood. Nevertheless, because pulsars are such stable clocks, they netted the discovery of the first extrasolar planets[1] and are the prime laboratories for the study of General Relativity in the weak and strong gravity regimes.[2] Quasars, also complex and diverse beasts, are used to probe the chemical evolution and large-scale structure of the Universe. So we may indeed take it as a sign of maturation that GRBs have arrived into the noble ranks of the Superlative (Albeit Enigmatic) Probes of the Universe. This chapter is devoted to the "killer apps"[3] of GRBs.

The unassailable fact about GRBs that makes them such great probes is that they are fantastically bright and so can be seen to the farthest reaches of the observable Universe. As we shall see, from a practical standpoint, the uniqueness as powerful probes stems from the observation that afterglows are apparently featureless (or at least simply described) and that afterglows eventually disappear from view. For some applications, the inference that the events occur in and around where star formation happens is also critical. As beacons to interesting and far-off places in the darkness, GRBs are indeed marvelous lighthouses (figure 6.1).

6.1 Studies of Gas, Dust, and Galaxies

In the context of understanding the nature of GRBs, the importance of the inferences of gas and dust properties along the line of sight was discussed in §3.1.2 and §4.1.1. But what impact do such inferences have on our understanding of gas and dust in the Universe? We noted previously (§4.1.1) that metallicity can be inferred using the absorption fingerprints of different atomic species. Metallicity influences the efficiency of mass loss in stars and thus how (massive) stars end their life.* Metallicity may also affect the distribution of masses of stars at birth (the so-called initial mass function). The measurement of metallicity serves as an instantaneous snapshot of the cumulative effects of chemical enrichment in a galaxy caused by generation after generation of stars expelling their nuclear waste as they die. More massive galaxies not only produce more stars over time but also tend to retain the metals expelled during supernova explosions. This leads to an expected "mass-metallicity" relationship and, since galaxy luminosities generally correlate with mass, a similar "luminosity-metallicity" relationship. Measuring these relations and how they evolve with cosmic time thus

---

*Higher metallicity generally makes for more efficient mass loss.

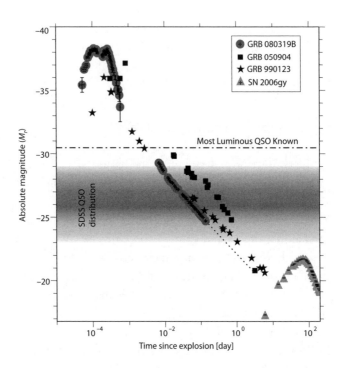

Figure 6.1. Bright gamma-ray bursts compared to other bright lighthouses of the Universe (quasars [QSOs] and supernovae). Plotted is the absolute optical magnitude (M_r) versus time since explosion. Recall that magnitudes are a logarithmic scale where more negative values imply a brighter source. A typical galaxy has $M_r \approx -18$ to -22. SN 2006gy was one of the brightest supernovae ever observed. The brightest QSOs found in the Sloan Digital Sky Survey (SDSS) are a factor of ten thousand times brighter than SN 2006gy at peak. The GRBs shown were about a factor of one thousand brighter at peak than the brightest quasar. From J. S. Bloom et al., *ApJ* **691**, 723 (2009).

provides a fundamental view of star formation and galaxy evolution.

Since the discovery and follow-up of GRBs depend little on the host galaxy properties, metallicity measurements using afterglows afford the study of a population of galaxies that were not selected because of their brightness. Thus, GRBs can provide a complementary picture[4] at the faint end of the luminosity function of galaxies. Initial results are intriguing: GRB hosts at redshifts above $z = 2$ appear to be more metal poor than predicted by simple prescriptions for chemical enrichment in galaxies based on their brightness. This suggests either an enhanced efficiency of star formation in dwarf galaxies or that SNe winds are more efficient at carrying away metals from galaxies.[5]

Broadband model fits to UVOIR afterglow data reveal a broad distribution of absorption depths due to dust. Most well-studied afterglows exhibit little to no extinction.* Those that do can be used to infer the properties of the dust in other galaxies. What is measured is the degree to which light at some wavelengths is absorbed when compared to the absorption of light at other wavelengths; this selective extinction is a direct manifestation of the properties of the dust grains themselves. This distribution,

---

*This reflects both intrinsic and extrinsic selection effects. Very extinguished afterglows—externally dimmed at UVOIR wavebands—are less likely to be detected than those afterglows that have not been absorbed as much; we will explore these so-called "dark bursts" in §6.2. Such high degrees of absorption in afterglows are fairly rare, not too surprising given that the majority of the light coming from ordinary stars in the Universe is also not heavily obscured by dust. In the cases where the immediate environment of the GRB is itself dusty, the afterglow may destroy enough dust in the first few seconds to erase the signs of the enshrouding dust.

in turn, derives from the nature of the dust factories that pollute the GRB sightlines. In the local universe (perhaps even in the last ten billion years or so), the distribution of dust is thought to be dominated by material from the late stages of stars once like our Sun (called "asymptotic giant branch" [AGB] stars). We strive to understand the nature of dust because it affects all our UVOIR measurements of objects and events outside the galaxy.* Despite the centrality of dust in extragalactic studies, it is important to note that measuring dust beyond the local universe is exceedingly difficult because it requires knowledge of the intrinsic spectrum of the background light source. Quasars and galaxies have complicated intrinsic spectra but GRBs do not, and in this sense, GRB afterglows are potentially more useful for measuring dust. Indeed, the measurement of extragalactic dust using GRB afterglows is becoming an important cottage industry.

In many sightlines, a selective extinction profile commonly seen in the Small Magellanic Cloud (SMC) is preferred over other locally observed profiles (see also §4.1.1). Since the SMC shares many of the characteristics (similar low mass, low luminosity, and active star formation) of GRB hosts, this is not too surprising. A common Milky Way ultraviolet absorption feature (called the "2175 Å bump"[6]) is not seen in the SMC and, likewise, only seldom observed in GRB afterglows. Interestingly, in the well-observed afterglow of GRB 071025 there is good

---

*Dust, for instance, is a major contributor to the uncertainty of cosmological parameters derived from standard-candle measurements. See §6.5 for an expanded discussion.

statistical evidence that none of the local flavors of dust can account for the inferred extinction; instead, a (theoretical) dust profile originating from SNe provides a much better match to the data (see figure 6.2). This is the best-fit dust profile using a GRB beyond $z = 3$ and constitutes some of the best evidence that dust was different in the early universe.[7] One explanation for this is straightforward: at the high redshift of some GRBs, there was not enough time in the Universe for an appreciable number of stars comparable to the mass of the Sun to evolve to the AGB phase. So at that time the only major contributors of dust were supernovae, not the evolved stars that are the dominant polluters of today's universe.

GRB afterglows also penetrate galaxies between us and host galaxies. These "foreground" galaxies are completely unrelated to the GRB and any special ingredients (e.g., heavy-element metals) in the host galaxy cauldron that might enhance the rate of GRB production. Like the population of innocent bystanders caught in the crossfire of street fights, GRB-selected foreground galaxies offer up a more random sampling—and thus a more pristine view— of the distribution of galaxies. We first "see" these galaxies in absorption, typically in a common low-ionization metal species of magnesium (Mg II). This ion tends to be both blown to large distances from galaxies (upward of 50 kpc), and it is produced in such large quantities that it leaves an indelible absorption mark on the light that passes through it.[8] After the afterglow fades we can image the region around the GRB and study the properties of in-tervening galaxies. Not surprisingly this has given us an unprecedented window on the properties of galaxies on the

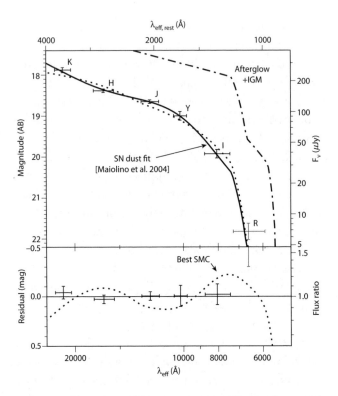

Figure 6.2. Inference of an SN-dominated dust profile in GRB 071025. The top panel shows the observed data in the optical and near-infrared as well as the model afterglow spectrum attenuated by neutral hydrogen in the intergalactic medium (IGM; dot-dashed lines). That is, the dot-dashed line shows the GRB afterglow as it would appear without dust. The dotted line is the dust-attenuated afterglow using a model for SN-dominated dust (from R. Maiolino, R. Schneider, E. Oliva, S. Bianchi, A. Ferrara, F. Mannucci, M. Pedani, and M. Roca Sogorb, *Nature* **431**, 533 [2004]); this line is an excellent fit to the data. The bottom plot shows the "residuals" (differences between the data and the model afterglow plus dust),

(*Continued*)

faint end of the "galaxy luminosity function": these galaxies might have been detected in other imaging surveys, but they are systematically absent from spectroscopic redshift surveys (which generally target the brightest galaxies). This is a fairly new direction for the field. In the future, we can use several sightlines to infer generic properties about the structure and extent of the halos of gas around galaxies as a function of galaxy properties.[9]

6.2 The History of Star Formation

Understanding star formation (SF) in the Universe is a central pursuit of observational cosmology. At a fundamental level, it helps us understand how we got where we are today: where the elements on Earth came from, why our Sun has the composition it has, etc. But SF also puts into context the diversity of galaxy structures we see and plays an important role in how galaxies evolve with time. The cumulative number of stars in a given galaxy can be inferred by mass and light measurements, albeit with some uncertainties. The *instantaneous*[10] rate of star formation is measured by a variety of techniques. At optical wavelengths, we measure the intensity of certain emission lines, which gives us some census of hot, big,

---

Figure 6.2. (*Continued*) including a result from a much poorer fit using Small Magellanic Cloud (SMC) dust extinction. From D. A. Perley et al., *MNRAS* **406**, 2473 (2010).

and newly formed stars. This is not the cleanest picture of SF because we need to assume a ratio between the numbers of less massive stars we do not detect and with those massive stars we do. More importantly, like fireflies in a dense fog bank, heavily obscured star formation goes unseen (at visible wavebands) and uncounted. Since this obscured light will serve to warm the dust that blocks it, we might infer how much nascent starlight we are missing by studying the reemission of starlight energy by dust (at infrared and sub-millimeter wavelengths). Studies of such dust have suggested that upward of one-third of star formation in the distant universe may be enshrouded and thus be undercounted in optical studies. Last, studies at radio wavelengths witness the effects of star formation in the form of emission from supernova shocks. There are many model dependencies in turning a radio flux measurement into an instantaneous star-formation rate (SFR), but, since radio light effectively penetrates dust, the radio-measured SFR tends to be viewed as the most robust.

The gamma-rays and X-rays from GRBs also penetrate dust. And since we believe that long-duration GRBs mark the birthplace (and death site) of massive stars, we get to pinpoint the locations of SF irrespective of whatever dust might get in our way. If some fraction of star formation is truly obscured by dust, then we might expect a similar fraction of GRBs to show evidence for dust extinction in the UVOIR afterglow. In the extreme case of opaque clouds of dust, the UVOIR afterglow might be completely extinguished. In the less extreme case, large amounts of dust might be relatively transparent to infrared light but effectively block ultraviolet light. Therefore, by measuring the distribution of concealed/obscured and unobserved afterglows, we might hope to measure directly the obscuration of *all* star formation in the Universe.

Faint or missing UVOIR afterglows are numerous— recall that roughly half of all Swift afterglows have no UVOIR afterglow detection. Some of the deficit can be chalked up to observational obstacles, such as when a GRB is detected near the Sun (even in this day and age we UVOIR astronomers still need darkness to do our business!) or if inclement weather hampers quick follow-up by telescopes with enough sensitivity to discover afterglows. GRBs detected during major sporting events, such as the World Cup of Soccer, or during August vacation recess for most of the world's GRB astronomers, tend to get less attention. But even some GRBs that get adequate rapid and deep follow-up have faint or undetected UVOIR afterglows. This population is referred to as "dark bursts." By connecting the brightness of the UVOIR afterglow to the brightness of the observed X-ray afterglow at some fixed time, we can ensure that the overall faintness of a GRB afterglow does not admit the event into the dark-burst club. Páll Jakobsson (University of Iceland) and collaborators proposed an operational definition of dark bursts that required the optical afterglow to be fainter than the most liberal synchrotron extrapolation from the X-ray data.[11]

In one homogeneous sample of thirty rapidly observed GRBs with a robotic 60-inch telescope at Palomar Mountain, fifteen events matched this dark-burst criterion. Of the entire sample, twenty-two (73 percent) had faint optical afterglows detected. In a follow-up study of the sample, my group at UC Berkeley showed that all but one event[12] had a plausible host galaxy detection (or optical afterglow detection). This, along with the observation that some of those hosts appear red (as if extinguished by dust throughout the galaxy), implies that ~25 percent of (high-mass) star formation in the Universe may be strongly obscured by dust.* This obscured fraction may indeed be an underestimate since, if the dust is near to the GRB explosion site, the light of the afterglow may quickly destroy the enshrouding dust; GRBs in such a configuration can thus claw themselves out of obscurity to join the UVOIR-detected group. The most extreme example of this is GRB 080607 which was observed to be an incredibly red afterglow, likely embedded in a thick blanket of dust (see also §4.1.1).

Dust issues aside, measuring the rate at which GRBs occur in the Universe as a function of redshift is a promising proxy for a measurement of the universal star-formation rate. There are some important caveats however. First, it appears that, as moss preferentially grows on the northern side of rocks and trees,[13] GRB progenitors prefer low metallicity (§4.1.2), implying that they serve as biased tracers of SF. Second, our ability to detect the GRB itself depends strongly on its intrinsic brightness as well as its distance from us. Third, and perhaps most important, our

---

*There are some alternative possibilities for the nondetection of optical after-glows that we considered in our study but did not favor, such as different afterglow spectra with intrinsically fainter optical emission. See D. A. Perley et al., *AJ* **138**, 1690 (2009).

ability to obtain a redshift of a GRB depends crucially on the ability to detect its optical afterglow. This is especially true beyond a redshift of 2–3 where emission-line redshifts from host galaxies become exceedingly "expensive" (viz., time expended on big telescopes). Despite these challenges, by modeling the observed properties of the GRB population, both those with redshift and those without, we can infer the rate of GRBs as a function of redshift and compare it to the rate of star formation as inferred by other means. Figure 6.3 shows this inferred rate. We see a rapid rise of the GRB rate from $z = 0$ (now) to $z = 1$, and then a sharp decline after $z = 4$. This is qualitatively similar to the inferred SFR except that the metallicity bias suppresses the GRB rate at low redshift relative to the universal SFR.

6.3 Cosmic Dawn: Measuring Reionization and the First Objects in the Universe

If we appeal only to the finite sensitivity of detectors, we might (on purely demographic grounds) expect that our detailed knowledge of the constituents of the Universe decreases monotonically as we move to higher and higher redshift. This is a reasonable first guess, but (unintuitively) it breaks down at the very highest redshift: we actually know a great deal about the Universe and its constituents around a redshift of $z = 1100$, about three hundred thousand years after the Big Bang. This is the redshift at which the light of the cosmic microwave background (CMB) originates, the epoch when the evolution of light and ordinary matter decoupled from their

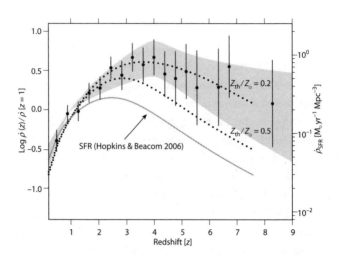

Figure 6.3. The inferred rate of long-duration GRBs, $\dot{\rho}(z)$, relative to the star-formation rate (SFR) in the Universe ($\dot{\rho}_{SFR}$), normalized to the rates at $z = 1$. The light gray curve shows a smooth fit to the observed SFR data from various studies of galaxies. The data points with error bars (and associated gray region) show the inferred GRB rate from a fit to the Swift data. The dotted lines show what a metallicity-biased SFR might look like, created by imposing a metallicity threshold cut ($Z_{th}/Z_{\odot}$ no more than 20 percent and 50 percent) on galaxies that can create stars. Remarkably, this metallicity-biased SFR looks similar to the inferred GRB rate, suggesting a strong correlation between metallicity and GRB production. Adapted from N. R. Butler, J. S. Bloom, and D. Poznanski, *ApJ* **711**, 495 (2010).

until-then-entwined existence; beyond this redshift is an opaque cloud through which we can never peer with electromagnetic eyes.[14] After that epoch, as the Universe cooled, neutral atoms of hydrogen could persist for the first time. Decades of study of the CMB and clouds of primordial gas at lower redshift have given us precise measurements of the distribution of matter at that time. Relative to the diverse and extreme clumpiness we see around us today,[15] the Universe then was a rather simple and placid soup of mostly hydrogen and helium atoms. Importantly, CMB measurements give us a precise (few percent uncertainty) measurement of the time of this epoch since the Big Bang and the distance to the "surface" of the CMB-emitting cloud.

Our current interests lie at redshifts and distances intermediate to the nearby and CMB bookends. At redshifts less than 1,100 and higher than about ~10, we know that the first clouds of primordial gas began to collapse, that the first stars were formed, that the first heavy elements were synthesized, and that the first galaxies were assembled. We also know that the Universe underwent a fundamental phase transition, changing from 100 percent neutral hydrogen (at $z < 1100$) to where almost all hydrogen was a fully ionized plasma by $z = 5$ (and continues to be to this day).* That this "epoch of reionization" happens to coincide with the formation of the first stars and structures might suggest a causal connection, but this is far from certain. One of the holy grails in observational cosmology today is to measure the progression of ionization as a

---

*We seek to measure $\bar{\chi}_H(z)$, the average neutral fractional of hydrogen at a given redshift. The value $\bar{\chi}_H(z)$ is the ratio of the number density of neutral hydrogen atoms divided by the number density of neutral hydrogen atoms plus protons. At $z > 1100$, $\bar{\chi}_H(z) = 0$. At redshifts just less than $z = 1100$, $\bar{\chi}_H(z) = 1$ and likely remains near unity until around redshift 20–10 when it plummets to near zero during the epoch of reionization. At $z < 5$, $\bar{\chi}_H(z)$ is very small ($< 10^{-5}$).

function of time and to uncover the source (or sources) of reionization. This epoch in the Universe (about 100–800 million years after the Big Bang) is sometimes referred to as the Dark Ages, capturing both our lack of understanding of that time (figuratively) and the relative lack of stars that shine during it (literally).* *In situ* measurements of objects and events beyond $z = 7$ must then be the true path to enlightenment (again both figuratively and literally). This ability to push observations beyond $z = 7$ is one of the great hopes for GRBs as probes.

The extreme energy release in GRBs also leads to the extreme brightness of afterglows across the electromagnetic spectrum. As figure 6.1 demonstrates, optical afterglows can be thousands of times brighter than the brightest quasar in the Universe and millions of times brighter than the brightest supernovae and galaxies. Importantly, GRB afterglows also stay brighter than most quasars for hours after the GRB trigger. Therefore, we not only expect to see GRBs from very far away, but we also might expect to see them *farther* away than other bright sources or events. Allowing ourselves to denigrate the cosmic competition for a moment, we might also note that the intrinsic brightness of the typical quasar and galaxy appears to be decreasing at higher redshift, whereas there is no good evidence to suggest that GRBs are also getting fainter at high redshift.[†] Even more fundamental is the recognition

---

*However, this is also the time when the Universe as we know it begins to blossom; so a more rosy view of this epoch is captured in the name "Cosmic Dawn."

[†]This makes intuitive sense if you believe that most long-duration GRBs are due to the death of massive stars (at a qualitative level and putting aside the metallicity

that the Universe appears to create the largest (and/or most massive) gravitationally bound structures at later times.[16] Therefore, the progenitors of GRBs and SNe (that is, stars), were likely formed before big galaxies (large collections of stars), which likely formed before the first massive black holes (perhaps a collection of stellar remnants) that power quasars. This uniqueness in space and time is probably only important when considering measurements at redshifts beyond $z \sim 15$. Nevertheless, taking off our cheerleader hat, we must admit that GRBs at high redshift are exceedingly rare. As figure 4.2 implies, there is probably only 1–2 bright GRBs per year at the Swift detection threshold beyond a redshift of the highest redshift quasar ($z = 6.5$ in 2010). So while observations of high-redshift GRBs are the hottest and the most informative tickets in town, the performance is about as frequent as the annual Academy Awards. Even more unfortunate, show times vary, and because of earthly constraints (like rain and daytime hours) we often will not know that there is a spectacular event unfolding until it is too late.

How do we recognize high-redshift events if their redshift is not encoded in observations of the bursts themselves? The simplest answer is that we get a spectrum of the afterglow and look for telltale absorption lines. But the practical answer is more complicated: since we need to decide how to allocate precious spectroscopic resources on large-aperture telescopes, we not only have to discover the

---

issues, why should a GRB progenitor care substantially about what time it is since the Big Bang?).

precise position of the afterglow quickly,[17] but we also have to gain some confidence that it is a high-redshift candidate.

The good news is that Nature is rather kind on this front. Despite being ionized at some degree, the Universe at high redshift has enough neutral atoms of hydrogen to selectively block out light at a specific wavelength (1216 Å) corresponding to the electronic transition of the $n = 1 \rightarrow 2$ state.* As a photon with this specific wavelength passes by a hydrogen atom in the $n = 1$ state, it has a high probability of being absorbed, causing an excitation of the atom[†] and the removal of that photon from the light train on its way to Earth. Now, since the Universe is expanding, atoms of neutral hydrogen existing at different places (and times) will see the photons more and more redshifted the farther away they are from the event. So while the atoms block the same wavelength of light as they see it, they are effectively blocking different wavelengths of light as viewed by us. Clouds of neutral hydrogen are so numerous in the $z > 5$ universe that we experience an almost complete blocking of light blueward of the Lyman-α transition shifted by the redshift of the GRB. More precisely, if we look for optical or ultraviolet afterglow at wavelengths

$$\lambda_{\text{observed}} < (1 + z) \times 1216 \,\text{Å}, \qquad (6.1)$$

we should not find it. At $z = 7$, equation 6.1 suggests that we should witness a dramatic fall off (the so-called

---

*This is the Lyman-α transition, from the ground state to the first excited state.
[†]The atom will eventually reemit the energy, in the form of a photon, as it falls from the $n = 2$ to $n = 1$ state. Since it has a very low chance of reemitting the photon in the same direction it was heading before it was absorbed, the incoming beam of light at that wavelength is effectively extinguished.

"Lyman-α break") blueward of 9728 Å. This wavelength corresponds to the approximate transition between optical and near-infrared wavebands. So in this case we "look" for a bright infrared afterglow without a detectable optical afterglow. Recall that a lack of (or very faint) optical afterglow essentially defines the dark-burst sample (§6.2), and since most dark bursts are dust-extinguished, infrared observations are essential for distinguishing dark bursts (which would appear red at in the infrared bands) from truly high-redshift events (which would appear relatively blue at infrared bands). Importantly, the infrared observations must be made around the same time as the optical observations to minimize the effect of a variable light curve on the inferred colors.

Our prowess at photometrically identifying high-redshift GRBs continues to grow. Access to rapid infrared imaging has improved on both small- and large-aperture telescopes, and new instruments[18] provide simultaneous optical-through-infrared photometric imaging. By 2010 at least four events appeared to have *photometric redshifts* beyond $z > 6$, and two of these had spectroscopic confirmation of the inferred photometric redshift. GRB 050904 was at a redshift of $z = 6.29$, and GRB 090423 was at a redshift of $z = 8.2$, making it the most distant spectroscopically confirmed source/event in the Universe at the time (see figure 6.4). GRB 090423 occurred just $\sim$630 million years after the Big Bang, when the Universe was less than 5 percent its present age and eight hundred times more dense than now.

Obviously, the discovery of high-redshift GRBs confirms the long-held suspicion that the progenitors of GRBs

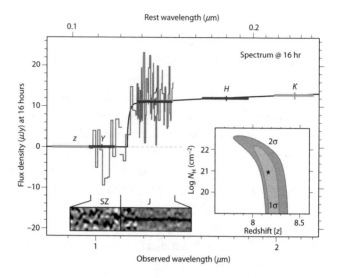

Figure 6.4. Spectrum of GRB 090423 about sixteen hours after the trigger. The dramatic drop-off of flux density blueward of about 1.1 μm is due to the collective opacity of neutral hydrogen clouds in the intergalactic medium along the line of sight. Letters ("z," "H," "SZ," etc.) are names of specific optical and infrared filters where imaging data were obtained. Inset shows a fit to the inferred redshift and hydrogen column density in the host galaxy of the GRB. Adapted from N. R. Tanvir et al., *Nature* **461**, 1254 (2009).

may be formed in the early universe when the typical metallicity of galaxies was lower than in today's universe.* This is an important statement about the progenitors of these events.

*The average metallicity at $z = 7$ was certainly lower than -2 (that is, one hundred times less metal abundance than our Sun) and probably closer to -3 (one thousand times less than the metal abundance of the Sun).

So what do we *do* with a high-redshift GRB? We saw in §6.2 that the incidence of high-redshift events helps us determine the rate of star formation in the Universe. But individual spectra are also potentially very powerful as diagnostics of the Universe itself. The "break" around redshifted Lyman α is not infinitely sharp: it progresses over tens of Ångströms, as dictated by the quantum mechanics of the electronic transition, effects of dust, the amount of neutral hydrogen surrounding the GRB in the host galaxy (N_H; see §4.1.2), and the ionization state of the Universe. By exploiting the powerlaw nature of the intrinsic afterglow spectrum, in principle, we should be able to determine the ionization state of the Universe at the time of the GRB. In practice, the two highest redshift GRBs with spectroscopic follow-up exhibited large N_H values, which washed out the measurement of χ_H. The hope is, based on low-redshift demographics of N_H values, that something like 10–20 percent of $z > 6$ GRBs will have low-enough N_H to yield a direct measurement of χ_H. One such measurement would be a great achievement and a set of such measurements at different redshifts would allow an unprecedented view of $\bar{\chi}_H(z)$, giving direct insight into the timing and origin of reionization.

Each well-observed GRB afterglow also allows us to constrain the gas-phase metals that it has encountered on its long journey to us. Along with a measurement of N_H, then, we can constrain the metallicity in surrounding subgalactic and galactic scales of the GRB (see also §4.1.2). A single absorption-line metallicity measurement beyond $z = 7$ would be novel and would certainly inform our understanding of the growth of stellar structure in the early universe. Perhaps more enticingly, we should be able to study the pattern of abundances of different metals with respect to each other. Just as the characteristics of dust may be shaped by the types of environments in which it was created, so too might abundance patterns reveal the origin of metals in the early universe. Beyond some redshifts the abundance pattern will be dominated by the ways in which the first stars (so-called "Population III" stars)[19] create and subsequently expel heavy elements. There is reasonably strong theoretical evidence to believe that Pop III stars are sufficiently different* from later generations in their lifecycles and death that we might witness a unique imprint in GRB absorption lines.

Irrespective of whether we have the sensitivity in our spectrographs to measure the effects of unique nuclear chemistry from Pop III stars in individual GRB afterglow spectra, it should be clear that GRBs offer a special service to high-redshift enthusiasts: they boisterously announce themselves and tell us where to look. Like any good lighthouse, high-redshift GRBs tell us where a special place in the Universe is. Months or even years later we can return to the position of GRBs and study the galaxy and even galaxy-cluster scales around where we know a high-redshift structure must exist. Knowing the GRB redshift ahead of time lets us tune up our imagers and spectrographs to get the most detailed physics out of the subsequent study.[20]

---

*In particular, they are thought to be much more massive than the average stars formed today.

6.4 Neutrinos, Gravitational Waves, and Cosmic Rays

We have so far concerned ourselves with the detection of light from GRBs. But of course messages from the heavenly bodies are delivered not just with electromagnetic light but via other forms of energy. The study of these "alternative" forms of radiation—which are very likely the manifestation of important physical processes in GRBs—should give us a very different perspective on the objects and events involved. The hope too is that we can use these other forms of emitted energy to test basic questions in physics.

6.4.1 Gravitational Waves

If indeed short bursts are due to the merger of compact objects (see §5.2), then in most scenarios, we expect that the GRB event would also be accompanied by a telltale signature of gravitational waves. Russell Hulse and Joseph Taylor demonstrated that the slow leakage of gravitational radiation from an NS–NS binary system could explain the gradual decay of its orbit. In about one billion years, the Hulse/Taylor binary will deteriorate to the point where the orbit will rapidly decay and the NSs will coalesce. As we have discussed (§5.2.1), in the last few milliseconds before the death plunge, the system will release a tremendous amount of energy in the form of gravitational waves.

The gravitational wave (GW) signal is oscillatory with a frequency comparable to the orbital time of the binary; so as the orbit decays, we expect the frequency of the signal to rise rapidly, like the chirping sound of birds.* And just as chirps from different species are distinctive, each GW merger event can be analyzed to reveal uniquely the properties of the merging objects (mass, size, perhaps even their internal structure).

Gravitational waves deform space as they propagate, and it is this deformation that gravitational-wave detectors measure. By 2010, there was no credible direct detection of gravitational waves from any astrophysical source despite the remarkably impressive sensitivities obtained by major facilities.[21] The main impediment is that the most sensitive facilities can only "hear" gravitational waves from NS–NS mergers out to ~10 Mpc, wherein the rates of NS–NS mergers are expected to be only about one event per ten years. Fortunately the Advanced LIGO project (starting science operation around 2015), which will improve the NS–NS sensitivity to 300 Mpc or so, should detect tens of events per year.[22]

Clearly a detection of gravitational waves emanating from the same time and place as a short-duration GRB would be a smoking gun for the progenitors of such events (§5.2.2). If the merger of an NS–NS or NS–BH binary caused the event, the chirp signal should be an unassailable confirmation of such configurations. In a broader context, the detection of a chirp from a GRB would be the first direct measurement of a gravitational-wave event and open

---

*Indeed, the GW signal of merger events is often referred to as a chirp signal. Interestingly, the last moments of NS–NS mergers produce chirp signals in the audible frequency range. You can "listen" to such mergers at a website maintained by Scott Hughes at MIT: http://gmunu.mit.edu/sounds/comparable_sounds/comparable_sounds.html.

up several new vistas on extreme physics. If, for instance, a neutron star is involved in the merger, then analysis of the chirp signal would allow a sensitive probe of the internal structure of neutron stars and reveal the nature of matter in an important new regime of high pressure and high density.[23] It would serve not only as a confirmation of General Relativity but allow stringent tests of General Relativity in a new regime where gravity is strong.[24]

6.4.2 Cosmic Rays and Neutrinos

Cosmic rays—heavy charged particles like protons and nuclei—bombard satellites in space and smash into the atoms in the Earth's atmosphere. Most low-energy cosmic rays are produced in the outskirts of the Sun,[25] and observational evidence presented in the mid-1990s showed that high-energy cosmic rays come from the remnants of supernovae. Essentially, SNe act as gigantic particle accelerators: charged particles moving with the nonrelativistic flow of an SN get bounced around in a magnetic field and eventually achieve speeds near the speed of light.

The very highest energy cosmic rays inferred* cannot have been produced in SN shocks, so some set of astrophysical entities must be responsible for the extreme particle acceleration. The origin of these cosmic rays were a complete mystery for decades. In 1995, Eli Waxman (then at the Institute for Advanced Study, Princeton)

---

*Around 10^{20} eV, comparable to all the energy required to light at 10 W light bulb for one second! This is several orders of magnitude higher than the particle energies obtained in the Large Hadron Collider (LHC).

and Mario Vietri (Università di Roma) noted that the relativistic fireballs of GRBs would make ideal particle accelerators[26] and could easily produce cosmic rays with energies $>10^{20}$ eV. Moreover, the rate of high-energy cosmic-ray detection was reasonably consistent with the approximate rate that GRBs could produce such particles. While GRBs remain one of the most viable and tantalizing production mechanisms for high-energy cosmic rays, a direct association of a given particle with a given GRB is very unlikely.*

However, if the hypothesis about particle acceleration is correct, some of the high-energy charged particles produced in GRB fireballs will interact with the gamma rays of the event itself.[27] These interactions lead to the production of smaller-mass particles (like positrons) and high-energy neutrinos.† Neutrinos produced in the fireball should have energies around 10^{14} eV. Neutrinos might also be produced from the decay of fast-moving protons in the external reverse shock, leading to a brief "neutrino afterglow" (typical neutrino energies of 10^{18} eV).[28]

Unfortunately, neutrinos are exceedingly difficult to detect and identify with their source: though about ten billion Solar neutrinos pass through an area the size of a penny in one second, only twenty-four neutrinos from

---

*Magnetic fields in the space between galaxies are large enough to cause the charged particles to take a meandering path. There are two effects of this: (1) Most particles from a given GRB will be deflected away from us, and (2) the time delay due to the meandering between the GRB event and the particle event will be about ten million years. This is too long to wait!

†Neutrinos, light-weight elementary particles, interact only weakly with matter and fly through the atmosphere, you, and the Earth almost unnoticed. Most low-energy neutrinos are produced in the Sun as a by-product of the nuclear fusion that powers the star.

an astrophysical source other than the Sun have been identified (and all were from a nearby supernova called SN 1987A). The good news is that, unlike cosmic rays produced in GRBs, we expect neutrinos from a GRB event to arrive nearly simultaneously with the gamma rays. This simultaneity would allow GRB neutrinos to be distinguished from neutrinos that are unrelated to the GRB and thereby improves the sensitivity of detection. The South Pole experiment called IceCube (reaching full sensitivity by 2012) and the Japanese Experiment Module on the Extreme Universe Space Observatory (JEM-EUSO; to be launched in 2013) are both poised to find the first neutrinos from GRBs.

Just as the detection of gravitational waves would be a major vindication of some progenitor models, the discovery of GRB-produced neutrinos would also be a major triumph for the fireball and afterglow theory. And as with gravitational waves, GRB-neutrinos would open up new perspectives on basic physics. For example, since neutrinos are particles with mass and gamma rays are not, it might be possible to devise stringent tests of the notion that gravity acts the same on light and mass.*

6.5 Quantum Gravity and the Expansion of the Universe

The extravagant reality of a universe accelerating away from itself by unknown means was a wonderfully fitting

---

*This is the so-called weak equivalence principle, one of the tenets of General Relativity. See E. Waxman, *Royal Society of London Philosophical Transactions Series A* **365**, 1323 (2007) for a review.

close to the twentieth century. Just as Hulse and Taylor were awarded the Nobel Prize for the astrophysical confirmation of gravitational waves (a basic prediction of General Relativity), inferences of a dark energy force counteracting the attraction of gravity pointed to a future of scientific inquiry beyond Einstein's dreams.* Coupled with technical advances in string theory, in some observational solidification of inflation theory,[†] and in sizing up the contribution of dark matter, the twenty-first century stage was set to start grappling with physics decidedly outside our comfort zone.

The very heart of Relativity—the notion that the speed of light is constant—has been prodded by beyond-Einstein thinkers for decades. Some believe that on the smallest size scales (something like 10^{25} times smaller than a hydrogen atom!) empty space becomes grainy enough to affect the propagation of light. On these scales, the deterministic tenets of gravity break down, and the unwieldy uncertainty of quantum mechanics reigns supreme. One ramification of this is that photons of different energy may travel at speeds slightly different than what we now think of as the universal speed c. Over very long distances, then, two photons with different energies beginning a cosmic race at exactly the same time will arrive at different times.

---

*To be sure, as noted already in §5.1.3, Einstein *did* include a cosmological constant in some of his original General Relativistic (GR) solutions for the dynamics of the Universe. While a seemingly good parameterization of universal acceleration, the apparent existence of a nonzero cosmological constant does not tell us *why* it exists. The answer—the physical origin of acceleration—lies decidedly beyond GR.

[†]Inflation posits another acceleration period in the Universe, just after the Big Bang.

Some have suggested that in distant GRBs, the arrival time of high-energy (GeV and TeV) photons relative to low-energy photons places the tightest limits today on this so-called Lorentz Invariance Violation (LIV) effect.

Some of the best-timed events, such as GRB 090510, *do* show a delay of photon-arrival times, but these are attributed to unmodeled intrinsic effects at the source (that is, reflecting when the photons started their race). Since we know so little about the emission mechanisms of GRBs compared to how much we know about General Relativity, it seems that all departures from equal-arrival time, even if tantalizingly consistent with a single LIV theory,[29] will always be chalked up to a more mundane explanation (such as a tauntingly evil framework of intrinsic GRB physics that mimics an LIV effect some hope to see; or, even worse, a conspiratorial evolution of GRB physics with redshift that gives a seemingly exciting LIV result with many GRBs).

On such grounds, most scientists, I believe, would agree that the extraordinary evidence* for a positive detection of LIV might never be obtainable. But a more subtle concern, almost epistemological in nature, arises: if assumptions about the emission mechanism of GRB physics must also be invoked to constrain LIV, isn't extraordinary evidence required as well?[†] I claim yes, and I also claim that extraordinary evidence of this sort has not been presented thus

---

*Carl Sagan is famous for saying that extraordinary claims require extraordinary evidence.

†That is (more formally), fail to disprove the null hypothesis of different arrival times as due to a breakdown in General Relativity.

far. Since my thoughts on this subject do not necessarily capture a widely held consensus, a positive spin on this LIV controversy is that, in just this one respect, GRBs are nevertheless pushing the envelope as probes of ideas beyond standard physical theory.

The idea of using GRBs to measure the evolution and fate of the Universe is also an exciting, albeit controversial, direction. Unfortunately, measuring and understanding the evolution of the Universe ("cosmography") is like trying to decipher the ingredients in a cake by carefully watching how the dough rises in the oven. That is, the task is immensely challenging. Thankfully, however, working under the aegis of General Relativity, the parameters that encapsulate the contribution of these ingredients are both finite and, in principle, measurable. The main task is to determine observationally how the redshift of the source maps to its distance from us. This is typically done by using large sources/structures whose sizes we can infer by some basic physics ("standard rulers") or whose intrinsic brightnesses we can infer by observations of local analogs ("standard candles").

GRBs are clearly not standard candles, in that the luminosity and energy release from burst to burst can vary by orders of magnitude (e.g., figure 4.3). However, if other observable parameters, such as afterglow break times, E_{peak}, duration, and such, could be shown to correlate intrinsically with luminosity or energy, then the events could be "standardized," implying that a GRB redshift measurement could also be tied to its true distance from us. Perhaps the closest analogy to "standardization" would be if you could tell the wattage of a newly lit light bulb by observing how fast it turned on or by its color. There are indeed many such correlations with GRBs, such as the so-called Amati relation connecting the isotropic-equivalent energy release in gamma rays (E_{iso}) to the observed peak energy (E_{peak}). Many claims have been made that by using a correlation (or combining many of them) the luminosity distance of a GRB may be inferred independently of the redshift. This remains a controversial subject, however, because many others (including those in my group) have suggested that the correlations come about largely as detector-threshold effects and, therefore, hold little cosmographic power. Regardless, given the statistical uncertainty in the relationships and those endemic to individual events measurements, it is difficult to see now how long-duration GRB energetics studies could ever be competitive with more-precise cosmographic tools.

One promising avenue for the cosmographic utility of GRBs comes not from long-duration bursts but from gravitational-wave detections of short-duration GRBs. If a GW event is indeed discovered, the impact on our understanding of the progenitors would be profound (§5.2, §6.4.1); the cosmographic connection is more subtle but no less sublime. GW chirps encode the distance to the source but not the redshift of the source* (recall that both redshift and distance are required for a cosmographic measurement), yet GW sources are exceedingly difficult to pinpoint on the sky. This means that obtaining a redshift

---

*This is basically the opposite of what most astronomers are used to: redshift of an astronomical source is relatively easy to obtain, but distance is not easily encoded in our observations.

of the galaxy host is even more hopeless than for GRBs before the afterglow era. But if a GRB is also found coincident in time with a GW event and its position is consistent with a crude localization of the GW event, then the afterglow of the GRB can be used to identify the host galaxy and measure its redshift.

The best estimates are that with a dozen or so GW–GRB events we could measure the expansion rate of the Universe locally* to an accuracy of just a few percent, at least a factor of two better than currently obtained by other means.[30] This, in turn, would improve our understanding of the precise contribution of dark matter and dark energy to the dynamics of the Universe. Importantly, this measurement would rely only on an appeal to General Relativity (to map the chirp signal to a distance) and so would be an independent check of cosmographic measurements made by other means.

6.6 The Future of Gamma-Ray Bursts: At the Nexus of Physical and Astrophysical Inquiry

The pace of transformative breakthroughs seen in the early days of the afterglow era has slowed. Frameworks for understanding the high-energy events, afterglows, and progenitors have settled in. And while the internal-external shock model still has some troubling disconnects with the data, it is difficult to see how the basic understanding of the events (compactness of the energy source, relativistic

---

*Measuring an important cosmological parameter called "Hubble's Constant."

motion, association with stellar death) could be shown to be vastly wrong. To be sure, there will be new GRB-discovery satellites that will yield fainter events at a higher rate. Some events will confirm, clarify, and (perhaps) expand the notions we have about the diverse zoo of progenitors. Some will be oddities that cause us to question our basic inferences about the physics that gives rise to the events.

On many fronts, the equanimity of a maturing endeavor has replaced the precociousness and verve of basic discovery. And yet, why does the GRB field continue to feel so vivacious and so inviting? One answer might be that there are just so very many interconnected facets of the GRB puzzle that no one subfield may be considered truly solved until the entire picture is seen with greater clarity. Even once rock-solid inferences such as jet breaks appear more tenuous with the enlightening onslaught of more data. Another answer is that, for all the vagaries about the phenomenon itself, GRBs have proven themselves to be unique and useful probes in so many burning questions we have about the physical universe. Soon, events discovered beyond $z = 10$ will allow us to probe the first galaxies and stars, and more nearby GRBs might offer a powerful gateway to the dynamic gravitational-wave and neutrino universe. Indeed, as we savor the richness of GRBs themselves, it is difficult to point to any other phenomenon that touches not just on such a wide range of pursuits but on inquires that are so central to twenty-first-century astrophysics and physics.

NOTES

Preface

1. See http://www.ssl.berkeley.edu/ipn3/bibliogr.html for an updated list, maintained by Kevin Hurley at the Space Sciences Laboratory, Berkeley, CA.

Chapter 1. Introduction

1. Derived from the Spanish word *velar*, meaning "to guard" or "to watch over with vigilance."
2. S. Singer, *Proceedings of the IEEE*, **53** (12), 1935 (1965).
3. A suspicious event from September 22, 1979 was found to be unlikely of nuclear-detonation origin by a blue-ribbon panel of physicists. That report was declassified in 2006. See F. Press et al., *Ad Hoc Panel Report on the September 22 Event*. See http://www.gwu.edu/~nsarchiv/NSAEBB/NSAEBB190/09.pdf in particular and the archive of the Vela September 22 Event at the National Security Archive, http://www.gwu.edu/~nsarchiv/NSAEBB/NSAEBB190/1980.
4. J. Bonnell, *A Brief History of the Discovery of Cosmic Gamma-Ray Bursts*. http://antwrp.gsfc.nasa.gov/htmltest/jbonnell/www/grbhist.html (1995).

5. R. W. Klebesadel, I. B. Strong, and R. A. Olson, *ApJ (Letters)* **182**, L85 (1973).

6. Long retired, but still a participant in some scientific meetings, Ray Klebesadel amusingly refers to the events without the word *ray* ("gamma bursts") perhaps out of modesty.

7. The 1965 journal paper about the Vela program (S. Singer, *Proceedings of the IEEE*, **53** (12), 1935 [1965]), for instance, discussed only events from the Sun and Earth as possible cosmic contaminants to a nuclear-detonation signal.

8. That physical model was proposed by Sterling Colgate (*Canadian Journal of Physics* **46**, 476 [1968]). It involved the production of gamma rays in the outskirts of a supernova explosion. This is now not the preferred theory for the production of gamma rays in most GRBs we discover.

9. This is a quote from Klebesadel et al. We now know why the search for an SN counterpart failed: most GRBs detected occur at distances that are much farther away than the typical distances of SNe discovered in the 1960s and 1970s.

10. I. B. Strong, R. W. Klebesadel, and R. A. Olson, *ApJ* **188**, L1 (1974).

11. W. A. Wheaton et al., *ApJ* **185**, L57 (1973).

12. Angular area on the sky, Ω, has units of angular distance squared. For wide areas, Ω is given in steradians. There are 4π steradians on the celestial sphere and one steradian = 3,283 square degrees.

13. J.-L. Atteia et al., *ApJ Supplement Series* **64**, 305 (1987).

14. Using the rise time (or, more generally, *variability time*) to infer source size is the "light travel time" argument. Since the entire surface emitting photons cannot "know" precisely when to light up, the propagation of the information to turn on can travel only as fast as the speed of light. The surface

size that participates in the radiation cannot be larger than $l = \delta t \times c$, where δt is the timescale for appreciable change. If $\delta t \approx$ tens of milliseconds, then $l \approx 6,000$ km.

15. After a star explodes as an SN, the ejected material continues to propagate outward, first plowing through circumstellar material, then sweeping up the comparably less dense gas and dust that lies between stars (the *interstellar medium* [ISM]). Over time the remnant expands to many *light-years* in size. The accumulated material glows across the electromagnetic spectrum. Individual SN remnants are readily detected in the Milky Way and nearby galaxies.

16. The LMC is a small satellite of the Milky Way and, despite its small mass, is a prodigious factory of supernovae.

17. In the case of a spinning NS, there is an appreciable amount of "rotational" energy available related to the spin of the star. This energy is roughly proportional to the spin rate of the star squared, times the radius squared, times the mass.

18. A list of SGR candidates is currently maintained at http://www.physics.mcgill.ca/~pulsar/magnetar/main.html.

19. C. Kouveliotou et al., *Nature* **393**, 235 (1998).

20. 10^{14}–10^{16} Gauss, more than thousands of times most other pulsars observed. R. C. Duncan and C. Thompson, *ApJ* **392**, 9 (1992).

21. In the late 1910s, Harlow Shapley had argued that the Sun was not at the center of the Galaxy by noting that *globular clusters* appeared to be distributed around the constellation Sagittarius (see figure 1.3), which is indeed the direction of the recognized center of the Milky Way.

22. This is the one reason why it is colder in the winter: light from the Sun hits the Earth more obliquely, so over a given area and given time, less energy is absorbed. The intensity of light falls as the cosine of the angle between the plane of the detector and the incident angle.

23. All GRB detectors have "blind spots," where events coming from certain directions are less likely to be detected. Moreover, satellites in low-Earth orbit cannot detect a GRB when it occurs on the opposite side of the Earth. The curators of BATSE carefully kept track of the efficiency for detecting GRBs as a function of time and location on the sky. Just as a national census uses the statistics of replies to determine the total number of citizens, the BATSE efficiency map was used to reconstruct a blind-spot-free map of the rate of events over the entire sky.

24. Buttons were handed out before the Great Debate so that people could display their scenario preferences. One said "Gamma-Ray Bursts Are from the Galaxy." Another said "Gamma-Ray Bursts Are Extragalactic." Just to hedge the bets, another set was passed around that claimed "Gamma-Ray Bursts Are Other." I still have all three buttons.

25. That night, for my M. Phil thesis project at Cambridge University, Nial and I were engaged in a project to image the locations of well-positioned GRBs from the IPN. Getting images of the new well-localized GRB 970228 seemed like a good thing to do!

26. There was also another intervening gaseous system along the line of sight at a lower redshift.

27. See   http://www.pbs.org/newshour/forum/may98/bang_5-14.html/.

28. The redshift was a measly $z = 0.0085$, and the distance to that galaxy is about 39 *Megaparsec (Mpc)*.

29. There is no hard-and-fast distance that distinguishes a source from being in the "local universe" and being of "cosmological" origin. But here I will take a source beyond ~ 400 Mpc ($z \approx 0.1$) to be at a truly cosmological distance.

30. See S. E. Woosley and J. S. Bloom, *ARA&A* **44**, 507 (2006) for an extensive review.

31. Many physics and astronomy experiments, like BATSE, are acronyms, but Swift is not. The observatory was so named given its ability to *swiftly* repoint on new GRB positions—in less than one minute—with its sensitive X-ray and ultraviolet cameras. It was launched in November 2004.

32. J. S. Bloom et al., *ApJ* **638**, 354 (2006).

Chapter 2. Into the Belly of the Beast

1. Integral, a European mission launched in 2002, localizes about one GRB per month to an accuracy of a few arcminutes in radius.

2. This is named after the author David Band who introduced the functional form in D. Band et al., *ApJ*, **413**, 281 (1993). A small minority of GRBs appear to have an additional emission component at very high energies (above a few MeV extending to GeV energies) (K. Hurley et al., *Nature* **372**, 652 [1994]; see for review J. Granot et al., *GRB Theory in the Fermi Era*. arXiv/0905.2206 [2009]). There are also suggestions that some GRBs have two breaks in the spectra—the one usually seen in the gamma-ray *bandpass* (E_{peak}) and another in the X-ray bandpass.

3. E. E. Fenimore, J. J. M. in 't Zand, J. P. Norris, J. T. Bonnell, and R. J. Nemiroff, *ApJ* **448**, L101 (1995).

4. Our gamma-ray detectors are like our ears, sensitive to only a range of energy (or pitch). For an orchestra moving very fast away from us, the flute melody would be redshifted in frequency, sounding to have as low a pitch as a cello. The low pitches of the cello part might be redshifted outside of our sensitivity range, inaudible to us. Likewise, for GRBs

at high redshift, we detect high-energy light that has been redshifted into our sensitivity range, and we do not detect low-energy light that has between redshifted out of our sensitivity range.

5. See D. Lazzati and M. C. Begelman, *ApJ* **700**, L141 (2009) for a review of the observations and implications of polarization.

6. Just as the sky changes brightness throughout the day and night, there are times in the orbit of a satellite when the background brightness of the sky is much higher than at other times. This is due both to a combination of persistent astrophysical sources in the field of view of the detector and the changing flux of charged particles entrapped in the Earth's magnetic field.

7. These definitions are found in T. Sakamoto et al., *ApJ* **679**, 570 (2008).

8. And indeed individual pulses in a given event could be classified among these three groups.

9. E. P. Mazets et al., *Ap&SS* **80**, 3 (1981); J. P. Norris, T. L. Cline, U. D. Desai, and B. J. Teegarden, *Nature* **308**, 434 (1984).

10. C. Kouveliotou, C. A. Meegan, G. J. Fishman, N. P. Bhat, M. S. Briggs, T. M. Koshut, W. S. Paciesas, and G. N. Pendleton, *ApJ (Letters)* **413**, 101 (1993).

11. Think about the decision process before eating berries you find in the woods.

12. Radioactive material loses energy "exponentially," meaning that most of the energy loss happens very quickly after the material is synthesized; then the rate of loss decreases with time. The half life of a radioactive element dictates how fast the activity declines with time.

13. Correct: you heat up just a bit when you get an X-ray at the dentist office.

14. Blackbody-only fits are not statistically consistent with average GRB spectra. However, a combination of blackbody plus a nonthermal powerlaw has been shown to be consistent with some GRB spectra (e.g., A. Pe'er, F. Ryde, R. A. M. J. Wijers, P. Mészáros, and M. J. Rees, *ApJ* **664**, L1 [2007]), though "Band plus powerlaw" models also appear to fit GRB spectra as well (e.g., A. A. Abdo et al., *ApJ* **706**, L138 [2009]).

15. See E. E. Fenimore, R. I. Epstein, and C. Ho, *A&AS* **97**, 59 (1993).

16. M. Ruderman, *Annals of the New York Academy of Sciences* **262**, 164 (1975).

17. See R. Sari and T. Piran, *ApJ* **485**, 270 (1997) for an extended discussion of the implications of observed variability in the context of relativistic motion.

18. See for review, J. Granot et al., *Highlights from Fermi GRB Observations*. arXiv/1003.2452 (2010).

19. See n. 16 above.

20. Weakly interacting particles, like neutrinos, could escape this opaque fireball and carry away energy. Indeed, it is believed that more energy is carried by neutrinos than is confined in the photon/particle soup. We return to neutrinos in §6.4.2.

21. When the particle is at rest, $\Gamma = 1$ and this formula reduces to the well-known equation $E = mc^2$. The mass m is called the *restmass* of the particle. When the object is moving with low speeds ($v \ll c$), then $E \approx mc^2 + (1/2)mv^2$; that is, the total energy associated with the particle is the restmass plus the traditional value for the kinetic energy of a moving particle of mass m.

22. The most natural interactions at this scale would be Coulomb interactions (that is, via forces between charged particles), which causes scattering. This is the dominant

interaction that transforms kinetic energy into heat when a car smashes into a brick wall.

23. P. Mészáros and M. J. Rees, *MNRAS* **257**, 29P (1992).

24. M. J. Rees and P. Mészáros, *ApJ* **430**, L93 (1994).

25. This is the so-called Schwartzchild radius, which is not a physical surface like that of a rocky planet but instead represents an important boundary. Inside this radius, nothing can escape the black hole (not even light); outside the boundary, radiation from mass *accreting* into the black hole can be detected by distant observers. For a black hole of mass M that is not spinning, this radius is $R_s = 2GM/c^2$.

26. Following from M. Rees. See A. M. Beloborodov, in *American Institute of Physics Conference Series*, ed. M. Axelsson, vol. 1054, 51 (2008) and W. H. Lee and E. Ramirez-Ruiz, *New Journal of Physics* **9**, 17 (2007) for reviews.

27. B. D. Metzger, E. Quataert, and T. A. Thompson, *MNRAS* **385**, 1455 (2008).

28. R. Blandford and D. Eichler, *Physics Reports* **154** (1), 1 (1987) provide a general overview.

29. Light, of course, is an electromagnetic wave. To start the propagation of this wave (i.e., emit light), a changing magnetic field or a changing electric field must be produced. The simplest way to do this is to accelerate a charged particle. In fact, if you wave an empty soda can around, the free charges in the metal will be accelerated, and you will generate light. In this case, you would generate long-wavelength radio waves. If someone nearby had a sensitive-enough receiver, they could listen to your soda music.

Chapter 3. Afterglows

1. A. C. Fabian, V. Icke, and J. E. Pringle, *Ap&SS* **42**, 77 (1976) appear to have first used the term "afterglow" to describe delayed emission following a GRB (from a Galactic neutron star); see also D. Eichler and A. F. Cheng, *ApJ* **336**, 360 (1989); J. E. Grindlay and S. S. Murray, in *X-ray Astronomy in the 1980's*, ed. S. S. Holt (Greenbelt, MD: NASA, 1981), 349–366.

2. N. Gehrels, *Bulletin of the American Astronomical Society* **26**, 1332 (1994).

3. J. I. Katz, *ApJ (Letters)* **432**, L107 (1994); P. Mészáros and M. J. Rees, *ApJ (Letters)* **418**, L59 (1993); P. Mészáros, M. J. Rees, and H. Papathanassiou, *ApJ* **432**, 181 (1994).

4. B. Paczyński and J. E. Rhoads, *ApJ (Letters)* **418**, L5 (1993).

5. To maximize the error-box coverage in the least amount of time, searches were conducted at longer wavelengths where the effective field of view on the sky of radio receivers is larger. Unfortunately, as we later learned, the long-wavelength light of radio afterglows is systematically suppressed relative to shorter wavelengths due to a process called "synchrotron self-absorption."

6. P. Mészáros and M. J. Rees, *ApJ* **476**, 232 (1997).

7. B. J. McNamara, T. E. Harrison, and C. L. Williams, *ApJ* **452**, L25 (1995) provide a good overview of the counterpart searches before the beginning (1997) of the afterglow era.

8. This X-ray pulse was similar to the events detected by Ginga for years; such X-ray events in Ginga were considered to be part of the prompt emission.

9. See M. de Pasquale et al., *A&A* **455**, 813 (2006) for a review of BeppoSAX afterglow observations.

10. L. Piro et al., *ApJ* **623**, 314 (2005).

11. This is especially true in the first ~1 day after trigger. At later times, a powerlaw behavior is often allowed in fits to sparsely sampled X-ray light curves.

12. D. B. Fox et al., *Nature* **437**, 845 (2005).

13. See M. Sako, F. M. Harrison, and R. E. Rutledge, *ApJ* **623**, 973 (2005) for an excellent review and a systematic reanalysis of all available X-ray data before Swift.

14. V. Connaughton, *ApJ* **567**, 1028 (2002).

15. T. W. Giblin, J. van Paradijs, C. Kouveliotou, V. Connaughton, R. A. M. J. Wijers, M. S. Briggs, R. D. Preece, and G. J. Fishman, *ApJ* **524**, L47 (1999).

16. See, for example, P. Kumar and R. Barniol Duran, *External forward shock origin of high energy emission for three GBRs detected by Fermi*. arXiv/0910.5726 (2009).

17. With sufficiently rapid observations of arcminute-sized localizations, optical afterglows can be found independently of an X-ray afterglow discovery: it is of course possible to find optical afterglows without X-ray localizations, but the search, more laborious because a wider part of the sky must be searched, is not as consistently fruitful as when an X-ray afterglow is found.

18. The actual number of "dark bursts" is a matter of definition of what dark means. This is discussed more fully in §6.2.

19. This was the so-called naked eye GRB because the afterglow could have been seen with the unaided human eye. No one actually reported seeing this event by eye: only robotic telescopes recorded the very bright afterglow. Still, GRB 080319b smashed the distance record as the farthest object that *could* be seen with the naked eye.

20. C. H. Blake et al., *Nature* **435**, 181 (2005).

21. The early optical light curve morphologies are discussed in A. Panaitescu and W. T. Vestrand, *MNRAS* **387**, 497

(2008); A. Melandri et al., *ApJ* **686**, 1209 (2008); E. S. Rykoff et al., *ApJ* **702**, 489 (2009).

22. S. R. Oates et al., *MNRAS* **395**, 490 (2009).

23. In the Solar System, small dust grains less than 1 mm in size (called "meteoroids") entering the Earth's atmosphere at high velocities burn up and make streaks of light called "meteors." Single-molecule grains called "micrometeoroids" reflect and reemit sunlight; they are responsible for the Zodiacal light you can see at a dark observing site at night when the Moon is below the horizon. Dust grains in the very early days of the Solar System are thought to be the sites around which larger bodies are formed: comets, asteroids, moons, and planets. Shakespeare's Hamlet aptly called man and woman the "quintessence of dust." Attempting to understand this basic building block, the NASA mission Stardust collected dust from the tail of a comet and returned the samples back to Earth.

24. There is a wide range of values of α and β, even for different afterglows at the same stage in their evolution. See D. A. Kann, S. Klose, and A. Zeh, *ApJ* **641**, 993 (2006) (β values) and A. Zeh, S. Klose, and D. A. Kann, *ApJ* **637**, 889 (2006); S. R. Oates et al., *MNRAS* **395**, 490 (2009) (α values).

25. Dale Frail, the discover of GRB radio afterglows at the National Radio Astronomy Observatory (NRAO), has noted that most radio afterglows are detected within about a factor of ten of the sensitivity of the Very Large Array (VLA), whereas optical afterglows are generally detected several order of magnitude above the optical detection limits. The expanded VLA (eVLA) and the international facility Atacama Large Millimeter Array (ALMA) are expected to improve vastly the detection success at radio and millimeter wavebands, respectively.

26. See J. Goodman, *New Astronomy* **2**, 449 (1997).

27. See D. A. Frail, S. R. Kulkarni, L. Nicastro, M. Feroci, and G. B. Taylor, *Nature* **389**, 261 (1997).

28. G. B. Taylor, D. A. Frail, E. Berger, and S. R. Kulkarni, *ApJ* **609**, L1(2004).

29. P. Mészáros and M. J. Rees, *ApJ* **476**, 232 (1997).

30. R. A. M. J. Wijers et al., *MNRAS* **288**, L51 (1997).

31. The relationship between the observer time, Γ and r is $t_{\text{obs}} = r/2\Gamma^2 c$ under the assumption that Γ is not changing. Since we are considering what happens during a decelerating blastwave, the constant 2 in the denominator is not exact and should be replaced by ~ 4–6, depending on how the blastwave decelerates.

32. Carrying through the calculation keeping E_γ, the deceleration time t_{dec}, and n as free parameters, we see that

$$\Gamma_0 \propto \left(\frac{E_\gamma}{n} \right)^{1/8} t_{\text{dec}}^{-3/8}. \qquad \text{(N.1)}$$

So the longer the onset time for an afterglow, the lower the initial Lorentz factor.

33. With only a few facilities in the world with sufficient sensitivity to detect typical afterglows, radio, mm, and sub-mm data are the most difficult to obtain.

34. See E. Nakar and T. Piran, *ApJ* **619**, L147 (2005) for a summary.

35. Dale Frail noted at the early turn of the century the parallel of this concern with the worrying addition of "epicycles" to explain the improving data on planetary motion in the Ptolemaic model of the Solar System. Of course, the Keplerian explanation eventually emerged as a more elegant formalism. We hope the Johannes Kepler of GRB afterglows has been born, but it's possible he or she has not yet!

36. Note, however, that a wide range of p values from 1.8 to >3 have been inferred in some events. See M. C. Nysewander et al., *ApJ* **651**, 994 (2006); A. Panaitescu, *MNRAS* **362**, 921 (2005); A. Zeh, S. Klose, and D. A. Kann, *ApJ* **637**, 889 (2006).

37. A classical example of an afterglow that is well fit by the sort of nonuniform medium expected around massive stars is GRB 011121. See P. A. Price et al., *ApJ* **572**, L51 (2002).

38. See R. Sari, T. Piran, and R. Narayan, *ApJ (Letters)* **497**, L17 (1998).

39. See D. A. Frail et al., *ApJ (Letters)* **562**, L55 (2001) and a review by S. B. Cenko et al., *ApJ* **711**, 641 (2010).

40. E. Berger, S. R. Kulkarni, and D. A. Frail, *ApJ* **612**, 966 (2004).

41. See D. A. Frail, A. M. Soderberg, S. R. Kulkarni, E. Berger, S. Yost, D. W. Fox, and F. A. Harrison, *ApJ* **619**, 994 (2005).

42. See S. E. Woosley and J. S. Bloom, *ARA&A* **44**, 507 (2006) for a review.

Chapter 4. The Events In Context

1. The ISM is certainly clumpy but the typical scales of clumpiness are much larger than 10^{17} cm, the scale over which GRB blastwaves are propagated and afterglows are generated.

2. Interstellar dust grains are generally thought to be either carbon- or silicon-based molecules and exist in a range of sizes (from $<0.1\,\mu$m to ~0.1 mm). Dust is formed in the final stages of the lives of ordinary stars and also in the ejected material of supernova explosions. See §6.1.

3. Dust destruction has been suggested as the cause for the late turn-on of GRB 030418 (E. S. Rykoff et al., *ApJ* **601**, 1013 [2004]), but there are a number of alternative models. Dust destruction has specific observable signatures, namely the decrease in the opacity at optical wavebands and the commensurate dereddening of the afterglow. Interestingly, one event (GRB 061126) showed some evidence for "gray dust," where significant absorption existed in the UVOIR bands but without the expected reddening. Here the interpretation was that the GRB afterglow was effective at destroying small dust grains but did not manage to destroy the largest grains. If only the largest grains survive, they serve to block all wavelengths of light by the same amount. That is $\tau(\lambda) \approx$ constant.

4. Some high-velocity lines seen in absorption in GRB 021004 were initially taken to be due to fast-moving outflow from a massive-star progenitor. But some of the telltale atomic transitions that could not have persisted given the required intensity from the afterglow were seen. The more mundane interpretation, that the GRB 021004 afterglow absorption lines were actually due to a gas system physically disconnected from the host galaxy, seems to be much more favored. See H. Chen, J. X. Prochaska, E. Ramirez-Ruiz, J. S. Bloom, M. Dessauges-Zavadsky, and R. J. Foley, *ApJ* **663**, 420 (2007) for a review of the spectroscopic signatures in the CBM.

5. J. X. Prochaska et al., *ApJ* **691**, L27 (2009).

6. The inference came from observations of UV-excited lines of H_2. See Y. Sheffer, J. X. Prochaska, B. T. Draine, D. A. Perley, and J. S. Bloom, *ApJ* **701**, L63 (2009).

7. GRB 020813 (M. Dessauges-Zavadsky, H. Chen, J. X. Prochaska, J. S. Bloom, and A. J. Barth, *ApJ* **648**, L89 [2006]) and GRB 060418 (P. M. Vreeswijk et al.,

A&A **468**, 83 [2007]) are the first cases of temporal variation.

8. OK, perhaps that is a stretch!

9. Recall that SNe arise from the death of evolved stars, marking (in many scenarios) the end of life of stars more massive than about 8–10 $M_\odot$. For most of the lifetime of massive stars, they fuse light elements into heavier elements (i.e., metals). Upon explosion, those synthesized elements are ejected and strewn into the immediate environment of the supernova.

10. Metallicities for nearby GRB galaxies are almost exclusively required to be measured by looking at some atomic lines in emission (rather than absorption).

11. Think about the alternative, that only a few photons from a faint afterglow are collected on average per wavelength bin of our spectrograph system. In this scenario, in the presence of normal detection-related noise, if there were no collected photons from a certain wavelength, we could not be certain that the deficit was due to a statistical fluctuation or true absorption from intervening gas.

12. See M. J. Michałowski et al., *ApJ* **693**, 347 (2009) for a review of the environments inferred for low-redshift GRBs.

13. Determining, in practice, where a GRB occurs around a galaxy amounts to comparing an image of the afterglow and its surroundings while it is still bright with an image taken much later. Aligning (or "registering") the stars and galaxies in the two images is straightforward conceptually but is a bit of a black art when the highest precision is required.

14. In many cases, like with the first X-ray afterglow of a short-duration burst GRB 050509b, the location of the event could be whittled down to "only" tens of square arcseconds. Yet in the deepest images obtained by the Hubble Space Telescope, there can be dozens of faint galaxies inside error

boxes of such a size. In such cases, there is no way to know for sure which galaxy to "assign" as the true host galaxy. This is where probability theory comes in. We can treat each location as a fuzzy boundary of where the event could have occurred on the sky. Knowing also the density of galaxies on the sky (units of [galaxy arcsec^{-2}]) of a certain brightness, for any possible host we can use a series of statistical statements about the likelihood of that galaxy being physically associated. When we have spectroscopic redshifts, we gain more information, since galaxies at a lower redshift than the highest redshift system inferred in the afterglow spectrum cannot be physically associated.

15. Stars are born with a variety of masses, from ~0.08 $M_\odot$ to several tens (or more) solar masses. Even though there are far fewer massive stars than less massive stars, the light from those massive stars outshines the collective light of the less-massive stars. Since massive stars are also (generally) hotter, they appear more blue than less-massive stars. The result is that young clusters of stars will appear blue. However, massive stars die away more quickly than less-massive stars; so over time, clusters tend to appear more and more red.

16. See J. S. Bloom et al., *ApJ* **638**, 354 (2006).

17. You can play around with other combinations of redshift, time and distance using an on-line calculator hosted at the University of California, Los Angeles (UCLA): http://www.astro.ucla.edu/~wright/CosmoCalc.html.

18. We have argued (E. M. Levesque et al., *MNRAS* **401**, 963 [2009]) that the only GRB with observed $T_{90} < 2$ seconds and an absorption-line redshift ($z = 2.6$) has some non-negligible probability (~5%) of belonging to the long-duration class. Many high-redshift GRBs do appear to be "short" after a correction for cosmological time dilation is applied but their physical origin appears to be more likely from massive stars. See §5.4 for discussion.

19. Nearby, volume increases like distance cubed (as you might have suspected!), but its growth at larger distances is more complicated due to cosmological effects. Indeed, since the Universe has a finite age and light takes a finite time to reach us, there is a finite volume of the Universe we can see.

20. Emission-line redshifts of GRB host galaxies can be obtained, in principle, on a more relaxed schedule, well after the GRB has faded. But in practice, emission lines from $z > 1$ hosts become increasingly more difficult to detect at higher redshifts. Almost all spectroscopic redshifts obtained in the Swift era are afterglow absorption-line redshifts.

21. Here, we assume a nominal optical spectrograph with good sensitivity at wavelengths between $\lambda_1 = 4000$ Å and $\lambda_2 = 9000$ Å. The available redshift range given is determined using the redshift equation in the footnote on page 27.

22. See D. Guetta, T. Piran, and E. Waxman, *ApJ* **619**, 412 (2005) and references therein.

23. See I. Leonor, P. J. Sutton, R. Fray, G. Jones, S. Márka, and Z. Márka, Classical and Quantum Gravity **26**, 204017 (2009) for a summary.

Chapter 5. The Progenitors of Gamma-Ray Bursts

1. This is called "magnetic dipole radiation."
2. To be sure, a WD in a binary system will undergo changes if mass is transferred from its companion. In the most extreme case, it is thought that a Type Ia SN occurs when a WD accretes so much mass that electron degeneracy pressure is insufficient at holding up the star, causing it to explode.
3. S. E. Woosley, *ApJ* **405**, 273 (1993).
4. A. I. MacFadyen and S. E. Woosley, *ApJ* **524**, 262 (1999).

5. See S. E. Woosley and J. S. Bloom, *ARA&A* **44**, 507 (2006) for a review.

6. See K. Iwamoto et al., *Nature* **395**, 672 (1998), and the review article by Woosley and Bloom, *ARA&A* **44**, 507 (2006).

7. J. S. Bloom et al., *Nature* **401**, 453 (1999).

8. See R. Vanderspek et al., *ApJ* **617**, 1251 (2004), and references therein.

9. See the discovery papers (J. Hjorth et al., *Nature* **423**, 847 [2003]; K. Z. Stanek et al., *ApJ* **591**, L17 [2003]) of SN 2003dh associated with GRB 030329.

10. J. P. U. Fynbo et al., *Nature* **444**, 1047 (2006).

11. With so many GRBs observed, we certainly expect some number of such coincidences. The main challenge is in determining what the chance of such alignments might be for a given set of observations. See B. E. Cobb and C. D. Bailyn, *ApJ* **677**, 1157 (2008) for review.

12. And there may be more of them, such as GRB 051109b. See D. A. Perley, R. J. Foley, J. S. Bloom, and N. R. Butler, *GRB 051109B: Bright Spiral Host Galaxy at Low Redshift*. GCN Circular 5387 (2006).

13. This is akin to a fast-spinning ice skater, with arms tucked in, suddenly throwing his or her arms out wide. He or she will slow up quickly because of the mass placed at larger distances from the center of rotation.

14. See some of the earlier papers advancing this idea: D. Eichler, M. Livio, T. Piran, and D. N. Schramm, *Nature* **340**, 126 (1989); R. Narayan, B. Paczyński, and T. Piran, *ApJ* **395**, L83 (1992).

15. There is a possibility of a fourth class, a quark star, more compact than a neutron star and supported by pressures at the subnucleon scale. There is no convincing evidence to date that such objects exist in nature.

16. For instance, the other star may start dumping mass onto the NS and create an accretion disk, visible at X-ray wavelengths.

17. V. Kalogera, R. Narayan, D. N. Spergel, and J. H. Taylor, *ApJ* **556**, 340 (2001). See also R. O'Shaughnessy, K. Belczynski, and V. Kalogera, *ApJ* **675**, 566 (2008) for review.

18. There is certainly a possibility that this could be a collapsar leading to a GRB. A second GRB could then follow millions of years later!

19. This is the so-called escape velocity that depends on the galaxy mass and the birthplace of the progenitors. For big spiral galaxies like the Milky Way, the escape velocities can be ~100–200 km/s. For puny ("dwarf") galaxies, it can be a few tens of kilometers per second.

20. This is not to suggest that such a life path is unromantic or unfulfilling!

21. See A. Panaitescu, P. Kumar, and R. Narayan, *ApJ* **561**, L171 (2001).

22. L.-X. Li and B. Paczyński, *ApJ (Letters)* **507**, L59 (1998).

23. B. D. Metzger et al., *MNRAS* **406**, 2650 (2010).

24. Short-burst afterglows, generally fainter than long bursts, are more difficult to localize in X-rays. Moreover, only ~25 percent of short bursts have optical afterglows (compared to ~50 percent of long bursts) making the subarcsecond localization of short bursts relatively rare. With "large" error regions of a few arcseconds, there can be several potential host galaxies, especially at faint depths.

25. I am not a fan of cats, especially ones running loose in a crowd.

26. See J. S. Bloom et al., *ApJ* **654**, 878 (2007).

27. D. A. Perley et al., *ApJ* **696**, 1871 (2009).

28. There is some disconfirming evidence as well. For instance, it appears that short-duration bursts have roughly the same ratio of X-ray afterglow luminosity to GRB energy as long-duration bursts. Since X-ray afterglow luminosity should scale as the ambient density to the one-half power, if the mergers were happening preferentially in more tenuous circumburst environments, we would expect this ratio in short-duration bursts to be less than in long-duration bursts. So at face value, we would conclude that, absent some un-modeled biases, the circumburst densities of long-duration and short-duration bursts are about the same. (The subtlety of this interpretation is that we are assuming that the prompt event arises from internal shocks and the afterglow from external shocks.)

29. See D. A. Perley et al., *ApJ* **696**, 1871 (2009) for a review.

30. GRB 050906 (A. J. Levan et al., *MNRAS* **384**, 541 [2008]) may be a local-universe short-duration event found by Swift, but its association with a nearby galaxy is tenuous.

31. The notion that magnetars could be produced from an older population of binary white dwarfs has been advanced. See A. J. Levan, G. W. Wynn, R. Chapman, M. B. Davies, A. R. King, R. S. Priddy, and N. R. Tanvir, *MNRAS* **368**, L1 (2006).

32. The origin of GRB 070601 has been advanced as due to a BH binary undergoing a specific transition in accretion phase, a new regime of flaring from a magnetar, and planets smashing into white dwarfs.

Chapter 6. Gamma-Ray Bursts as Probes of the Universe

1. A. Wolszczan and D. A. Frail, *Nature* **355**, 145 (1992).

2. Hulse and Taylor won a Nobel Prize in Physics (1993), using precise timing of a pulsar–NS system to establish

observational evidence for gravitational waves. See R. A. Hulse and J. H. Taylor, *ApJ* **195**, L51 (1975); J. H. Taylor and J. M. Weisberg, *ApJ* **345**, 434 (1989).

3. "Killer apps" is used colloquially here. While usually associated with popular and transformative computer applications, in this context I mean to capture the unique applications of GRBs in broader aspects of physics and astrophysics inquiry.

4. Most galaxy studies use emission-line diagnostics to determine metallicities, giving a view of the chemical enrichment in regions where stars are formed. But GRB afterglow measurements (especially from events in the distant universe) use absorption diagnostics, giving access to the enrichment history in the interstellar medium of GRB hosts. The difference in techniques is not thought to dominate the overall comparative trends seen between the two methods.

5. H. Chen et al., *ApJ* **691**, 152 (2009).

6. The physical origin of the 2175 Å bump is not well known.

7. While not the first suggestion (R. Maiolino, R. Schneider, E. Olivia, S. Bianchi, A. Ferrara, F. Mannucci, M. Pedani, and M. Roca Sogorb, *Nature* **431**, 533 [2004]) of "supernova-smoke"-dominated dust, given the simplicity of the underlying afterglow spectrum, this is the least assumption-dependent statement about such dust. D. A. Perley et al., *MNRAS* **406**, 2473 (2010).

8. Strangely, there appears to be a higher incidence of strong Mg II absorbers along GRB sightlines than along quasar sightlines; see G. E. Prochter et al., *ApJ* **648**, L93 (2006). It is not clear whether this tells us more about quasars and GRB emission or something more fundamental about the size and structure of Mg II clouds in the halos of galaxies.

9. Such studies have been done for decades with quasars but suffer from a bias due to the persistence of the quasar: faint

galaxies and galaxies near the line of sight are clearly missed in such studies.

10. The adjective "instantaneous" is used loosely here. On cosmological timescales of billions of years, a measurement of the rate of star formation over the past 10–50 million years is an appropriately short snapshot of the recent activities in a galaxy.

11. P. Jakobsson, J. Hjorth, J. P. U. Fynbo, D. Watson, K. Pedersen, G. Björnsson, and J. Gorosabel, *ApJ* **617**, L21 (2004).

12. GRB 080320. GRB 090423 also had no optical detection of a host galaxy. See §6.3.

13. This is a preference, not a hard-and-fast rule. Use this rule of thumb as a last resort if you are lost in the woods (better to use the stars as a compass!). Note that in the Southern Hemisphere, moss prefers to grow on the southern side of surfaces.

14. In principle, we could "see" to even higher redshift by observing gravitational waves and neutrinos from this early time in the Universe. However, detecting these primordial signals is exceedingly difficult.

15. A cup of your favorite energy drink is $\approx 10^{29}$ times more dense than the average density of the Universe today. It is left as an exercise to you to determine what fraction of the total types of atoms in that drink were formed in SNe explosions.

16. The technical term for this is "hierarchical structure formation," which, despite its many syllables, flows rather trippingly off the tongue at cocktail parties.

17. Most spectrographic setups these days require that we have precise knowledge (at the <1 arcsecond level) of the afterglow position. Spectrographs made of "integral field units"—contiguous groupings of small but independent

inputs each with an individual spectrum output (e.g., from bundles of fiber-optic cable)—would allow a less refined afterglow position to be studied spectroscopically. But few large telescopes today currently have such units to be deployed rapidly.

18. Such as the GROND camera, which can take images of GRB positions in seven filters simultaneously (J. Greiner et al., *PASP* **120**, 405 [2008]).

19. Think of Population III stars as the first generation of stars, formed from primordial distributions of elements (mostly hydrogen and helium). Population II is the second generation of stars, sufficiently polluted with first-generation metals. Population I are all the stars that are formed after Pop II stars have had a chance to pollute their surroundings.

20. For instance, we might want to get a constraint on the mass of the galaxy (or protogalaxy) hosting the GRB using a specific emission line whose observed wavelength we could predict knowing the redshift of the GRB. We might also look for telltale emission signatures of Pop III stars, such as emission from ionized helium.

21. The fractional space-time compression that is now possible to detect is comparable to the change of the distance from here to the nearest star by an amount about the thickness of a human hair. It seems as though this sensitivity is still not good enough.

22. J. Abadie et al., *Class. Quantum Grav.* **27**, 173001 (2010).

23. We have not yet been able to determine how matter behaves (the so-called equation of state) at the densities and pressures in the cores of NSs.

24. See A. Dietz, *Searches for Inspiral Gravitational waves Associated with Short Gamma-ray Bursts in LIGO's Fifth and Virgo's First Science Run.* arXiv/1006.3393 (2010) and references therein.

25. Though not moving very fast, they have enough energy to ionize atoms in the Earth's atmosphere, giving rise to the Aurora Borealis (Northern Hemisphere) and the Aurora Australis (Southern Hemisphere).

26. M. Vietri, *ApJ* **453**, 883 (1995); E. Waxman, *Physical Review Letters* **75**, 386 (1995).

27. E. Waxman and J. Bahcall, *Physical Review Letters* **78**, 2292 (1997).

28. E. Waxman and J. N. Bahcall, *ApJ* **541**, 707 (2000).

29. One big problem is that there is no single LIV theory to be tested. So tests of LIV are stabbing at some small part of an otherwise only weakly constrained parameter space.

30. S. Nissanke, S. A. Hughes, D. E. Holz, N. Dalal, and J. L. Sievers, *Exploring Short Gamma-ray Bursts as Gravitational-wave Standard Sirens*. arXiv/0904.1017 (2009).

SUGGESTIONS FOR FURTHER READING

Many of the primary works where ideas or discoveries were first presented are referenced in the Notes. The easiest path to getting those works in a readable form is through the SAO/NASA Astrophysics Data System (ADS; http://adsabs.harvard.edu), a digital portal for physics and astronomy publications. Access to the published versions of some newer works will require that your home institution (e.g., a university) has an account with that specific publisher. Older publications should be retrievable by everyone. In most cases, even if access is restricted to the published version, you should be able to retrieve a preprint draft of the paper through the arXiv e-print link at the ADS site.

What follows is a listing of some overview works, either review articles or related books. Limited previews of most books can be found at Google Books (http://books.google.com).

Historical Perspectives

- *Flash! The Hunt for the Biggest Explosions in the Universe*, Govert Schilling (2002; Cambridge University Press): a wonderful popular-level account of the GRB story, focused on the beginning of the afterglow era. *Biggest Bangs*, Jonathan Katz (2002; Oxford University Press) is also a popular-level book written by a practicing theorist in the field. These books focus on the history of GRBs and the early days of the afterglow era, but many of the most transformative discoveries and insights you read here occurred after these books were published.

- Articles from the Great Debate summarizing the case for the Galactic and extragalactic distance scales: "The Distance Scale to Gamma-Ray Bursts," D. Q. Lamb (1995; *Publications of the Astronomical Society of the Pacific*, Vol. 107, p. 1152) and "How Far Away Are Gamma-Ray Bursters?," Bohdan Paczyński (1995; *Publications of the Astronomical Society of the Pacific*, Vol. 107, p. 1167).

High-Energy Emission and Afterglows

- *Gamma-Ray Bursts: The Brightest Explosions in the Universe*, Gilbert Vedrenne and Jean-Luc Atteia (2009; Springer Praxis Books). An in-depth look at the theory and observations of GRBs and afterglows. This book covers a similar array of topics as do I but is geared more for the practitioner in the GRB field or graduate students.

- *Gamma-ray Bursts in the Afterglow Era: Proceedings of the International Workshop, Held in Rome, Italy, 17–20 October 2000*, edited by Enrico Costa, Filippo Frontera, and Jens Hjorth (2001; Springer). Not particularly current but a good technical snapshot of our understanding of the GRB phenomenon well into the afterglow era.

- *Gamma-Ray Bursts in the Swift Era*, Neil Gehrels, Enrico Ramirez-Ruiz, Derek B. Fox (2009; *Annual Review of Astronomy and Astrophysics*, vol. 47, pp. 567–617). A very good technical review of the current state of GRB research in 2009.

- *The Physics of Gamma-ray Bursts*, Tsvi Piran (2004; *Reviews of Modern Physics*, vol. 76, No. 4, pp. 1143–1210) and *Gamma-ray bursts*, Peter Mészáros (2006; *Reports on Progress in Physics*, vol. 69, No. 8, pp. 2259–2321). Two technical reviews, focused on the theory of GRBs and afterglows.

The Supernova Connection

- *Cosmic Catastrophes: Exploding Stars, Black Holes, and Mapping the Universe*, J. Craig Wheeler (2008; Cambridge University Press). A popular account of the life cycles of stars that go bang, with a focus on the origins of supernovae.
- *Supernovae and Gamma-ray Bursters: Volume 598 of Lecture Notes in Physics*, edited by Kurt Walter Weiler (2003; Springer). A collection of short technical articles by prominent astronomers covering supernovae and GRBs.
- *The Supernova Gamma-ray Burst Connection*, Stan Woosley and Joshua Bloom (2006; *Annual Review of Astronomy and Astrophysics*, vol. 44, No. 1, pp. 507–556). A technical overview of the observational and theoretical connection of GRBs, massive stars, and supernovae.

Universal Context

- *Cosmology*, Steven Weinberg (2008; Oxford University Press). A comprehensive, graduate-student-level overview of model cosmology from the Nobel Prize-winning author of the immensely popular *The First Three Minutes*.

- *The Extravagant Universe: Exploding Stars, Dark Energy, and the Accelerating Cosmos*, Robert P. Kirshner (2004; Princeton University Press). A humorous, popular-level account of modern cosmology and the discovery of an accelerating universe.

GLOSSARY

Absorption line: manifest as the paucity of light at a certain wavelength due the blocking of light by a specific atom or molecule. See *absorption-line spectroscopy*.

Absorption-line spectroscopy: The study of gas properties in the line of sight to distant sources by analyzing the selective blocking ("absorption") of light at certain wavelengths. Like distinct fingerprints, each ion has its own characteristic set of absorption lines that betray their existence. Often absorption-line spectroscopy is used to determine the *redshift* (and hence distance) of a cosmological source, since the characteristic lines of certain ions shifted redward in a spectrum can be readily identified.

Accretion: the process whereby material falls into/onto a massive body, potentially contributing to the growth of that body.

Accretion disk: a pancake-like region where matter flows toward a central object. The matter is generally fed from another massive body outside the accretion disk and makes its way toward the central object on a spiral-like orbit. In the process of *accretion*, some of the gravitational potential energy of the matter is liberated in the form of heat, allowing accretion disks to be very hot and luminous.

Afterglow: light detected across the electromagnetic spectrum after the GRB itself has ceased.

Ångström (Å): a unit of length equal to 10^{-8} cm.

Angular diameter distance: the effective distance such that the apparent size of an object decreases inversely proportional to its distance from an observer.

Arcminute: a measure of angular distance. One arcminute is 1/60th of a degree. The two headlights of a car about twelve miles away would appear to be separated by 1/3 arcminute; the typical human eye would be just incapable of discerning that there are two lights but instead would see the lights merged into one.

Arcsecond: a measure of angular distance, one arcsecond is 1/60th of an *arcminute*. It is approximately 1/206265th of a *radian*.

AU: Astronomical Unit; defined as the average distance between the Earth and the Sun and determined to be 1.496×10^8 km.

Background radiation: emission from diffuse and random places in the sky, effectively serving as noise that makes detection of a specific source or event more difficult. Background light is generally considered constant in brightness (for a given wavelength range) but may appear to change in time depending on the location of the satellite and the subsequent shielding of portions of the sky (e.g., by the Earth).

Bandpass: a certain range in wavelengths or energies over which an instrument is sensitive to light.

BAT: the Burst Alert Telescope, on board Swift.

BATSE: the Burst and Transient Source Experiment, on board *CGRO*.

BeppoSAX: the common name for the Italian-Dutch satellite, "Satellite per Astronomia a Raggi X."

Black hole (BH): the most dense astrophysical object known, characterized by both mass and spin. The object has no physical surface but has a characteristic radius from which neither matter nor light can escape.

Blackbody: a "perfect" emitter and absorber; it is a thermal radiator that has a spectrum entirely characterized by the temperature of the material.

Brightness distribution: the number of sources or events that are observed to be brighter than some *flux* level.

c: the speed of light; 2.99×10^{10} cm s^{-1}.

CBM: circumburst medium. The material surrounding the GRB explosion.

Celestial sphere: the apparent location of stars and galaxies on the sky. Since we cannot perceive the distances to far-away light sources (see *parallax*), all the sources appear to be at the same distance from us; it appears that the sources are projected onto a far-away sphere.

Center-of-momentum frame: in the context of colliding particles, the speed and direction one would need to travel in order
to see the total momentum (=mass × velocity for particles with
mass) of all the particles sum to zero.

CGRO: the Compton Gamma-Ray Observatory. *BATSE* was
one of the experiments on CGRO.

Column density: see N_H.

Comoving frame: in the context of an explosion, the point of
view of a set of observers who are traveling with the flow, at the
same outward expansion rate of the explosion. In a cosmological
setting, the point of view of a set of observers within or near to
the galaxy where the GRB occurs. In that case, such observers
do not see any *cosmological time-dilation* or *redshift* effects of the
event.

Compact object: a massive astrophysical body, such as a *black
hole* or *neutron star*, which is much more dense than the Sun
and is not supported against collapse by pressure associated with
normal nuclear fusion processes (as in the Sun).

Cooling break (v_c): a transition in the spectrum of a GRB afterglow between two different powerlaw regimes. At frequencies
less than the cooling break frequency (v_c), electrons responsible
for the emission are said to be "slowly cooling" (radiating much
less energy than their kinetic energy). Above v_c, fast-cooling
electrons are radiating an appreciable amount of their kinetic
energy.

Cosmic rays: any charged particle hitting a detector. When
the particles are moving slowly, they are usually associated with

nuclear decay by-products (e.g., from the material surrounding the detector). Fast-moving cosmic rays, with a high amount of energy per particle, are thought to be generated in astrophysical events (e.g., supernovae).

Cosmological time dilation: an effect of the expanding universe whereby events that occur at high *redshift* (z) appear to progress more slowly to observers on Earth than to observers nearby the event. Specifically, an event which takes time T to progress as viewed by us takes time $T/(1 + z)$ to occur "in the frame" of the event itself.

Counterpart: the generic term for a transient that is spatially and temporally consistent with the position of another object or transient, usually observed at a different wavelength.

Degeneracy pressure: a stiffness to material that originates from a quantum-mechanical phenomenon. At sufficiently high density, some particles, such as electrons and neutrons, cannot be packed together any closer than some critical distance. At this critical packing, the particles resist attempts to squeeze them together; this acts like a pressure that can strongly resist the crush of gravity.

Doppler shift: the change of the apparent frequency or wavelength of emitted light due to motion of the emitting source with respect to the observer. We are all familiar with the Doppler shift of soundwaves, having listened to the changing pitches of ambulances as they speed toward us and then away from us. Sources moving toward the observer will appear to emit shorter-wavelength signals (higher pitch) than if the source was not moving. Sources moving away from the observer will appear to emit longer-wavelength signals (lower pitch), resulting in an apparent *redshift*.

Dynamics: how a source changes physically in time. In the context of synchrotron blastwaves, it describes the radial evolution of the shock with time.

Ejecta: the material ejected in a supernova explosion or a GRB.

Electromagnetic event: temporal change of an astrophysical source that gives rise to light or *photons*, the carrier of electromagnetic energy. See also *electromagnetic spectrum*.

Electromagnetic spectrum: the full range of energy, wavelength, and frequency of light. Radio light occupies the smallest-energy, longest-wavelength, lowest-frequency portion of the electromagnetic spectrum. Gamma-ray light occupies the highest-energy, shortest-wavelength, highest-frequency of the electromagnetic spectrum. Other regions of the spectrum include visible ("optical"), infrared, ultraviolet, and X-ray light.

Electronvolt (eV): the energy required to move one electron across an electric potential of one volt. Numerically, this is 1.602×10^{-12} erg. For reference, the energy required to ionize a hydrogen atom in the ground state is 13.6 eV.

Error box: the location of where a GRB could have occurred on the sky. The uncertainty in this determination often makes the location possible anywhere within some polygon-like projection onto the *celestial sphere*.

External shock: a relativistic shock that occurs between material ejected by the central engine and material in the circumburst medium. External shocks are thought most likely to give rise to the afterglow emission.

Fermi: a high-energy gamma-ray NASA satellite mission launched in 2008. The on-board instruments can detect GRBs over the photon energy range of 150 keV to 300 GeV. The Large Area Telescope (LAT) has the best imaging *resolution*, and most LAT-detected GRBs have led to the detection of an afterglow.

Fireball: a hot mixture of (charged) particles, photons, and magnetic fields. The particles move fast with random motions. In the context of the early evolution of a GRB, the fireball carries the energy deposited near the central engine (§2.3) and expands outward radially.

Fluence: the energy collected per unit area. It is the integral over time of the *flux* from a source. Fluence is related to the total energy output of a source, and flux is related to the instantaneous brightness or luminosity of a source.

Flux: the energy collected per unit area per unit time. See also *fluence*.

Forward shock: the collisionless shock connecting the outflowing material with the circumburst medium. The forward shock is thought to be the site of the late-time afterglow emission.

γ-ray (gamma ray): wavelength regime of the electromagnetic spectrum shorter than X-rays; photon energies above $\sim$10,000 *eV*.

Globular cluster: a bound, roughly spherical swarm of millions of old and red stars; globular clusters also contain thousands of neutron stars and white dwarfs. There are a around two hundred globular clusters in and around the Milky Way.

Gpc: Gigaparsec; one billion parsecs, which is 3.085×10^{22} km.

Gravitational redshift: an apparent *Doppler shift* of light arising near anything that has mass. The amount of gravitational redshift, predicted by General Relativity, has been verified experimentally.

Gravitational waves (GWs): ripples, produced by accelerating masses, that deform space and time and are thought to travel at the speed of light.

HETE: High-Energy Transient Explorer.

HST: Hubble Space Telescope.

IGM: intergalactic medium.

Internal shock: a relativistic shock that occurs between material ejected by the central engine. Internal shocks are the interactions thought most likely to give rise to the prompt emission of GRBs.

Inverse Compton (IC) scattering: an interaction between a *photon* and (typically) an electron, whereby the *kinetic energy* of the electron is reduced and the energy of the outgoing photon is higher than the energy of the incoming photon. The direction of the electron's trajectory (as well as that of the photon) is altered by the interaction.

IPN: Interplanetary Network of Satellites.

IR: infrared; the wavelength regime between visible and the sub-millimeter. Near-IR observations span from about 10,000 Ångström to 26,000 Ångström (1–2.6 μm).

ISM: interstellar medium.

Isotropic/isotropy: the same in all directions. In the context of locations, isotropy implies that the typical distances between different GRBs is the same in all places on the sky. The measurement of the degree of isotropy (or anisotropy) is statistical in nature. In the context a given GRB, isotropy would imply that the energy released in all directions is the same. We do not believe that GRBs emit isotropically but instead are collimated.

Kinetic energy: the energy associated with the motion of a particle or body with mass; at low velocity (v), it is the familiar $\frac{1}{2}mv^2$.

kpc: kiloparsec; one thousand parsecs, which is 3.085×10^{16} km.

Light-year: the distance that light will travel in the vacuum of space in one year. This is about 9,436 billion kilometers or 0.306 *parsecs*.

LIGO: Laser Interferometer Gravitational-Wave Observatory.

LMC: Large Magellanic Cloud, a satellite galaxy of the Milky Way.

Lorentz factor (Γ): a term related to the velocity, v, of source as $1/\sqrt{1 - v^2/c^2}$. It is most useful to consider Γ instead of v when v is very close to the speed of light c.

Luminosity distance: the effective distance between two sources such that the $1/r^2$ law for the dimming of light is satisfied.

$M_\odot$:   the mass of the Sun; 1.99×10^{33} gm.

Magnetar: a highly magnetized *neutron star* thought to be the origin of *Soft Gamma-ray Repeaters* and possibly some extragalactic short-duration GRBs.

Magnitude: a measurement, in logarithmic scale, of the brightness of a UVOIR event. The "apparent magnitude" is what is measured, and the "absolute magnitude" is the brightness that would be measured of that object if observed from a distance of 10 *parsec*. The star Vega is defined to have an apparent magnitude = 0 at all wavelengths. Fainter sources have larger magnitudes such that, for every change of 2.5 magnitudes, a source is ten times fainter.

Metallicity: a measurement of the abundance of synthesized (heavy) elements relative to the abundances observed in the Sun. Given in dimensionless logarithmic units, a metallicity of -1 implies an enrichment of metals from the primordial abundances that is ten times less than in the Sun. Galaxies in the Universe two billion years after the Big Bang have typical metallicities in the range -3 to -2.

Mpc: Megaparsec; one million parsecs, which is 3.085×10^{19} km.

N_H: the column density of hydrogen (H), in units of cm^{-2}. Physically, this is the integral of the number density of hydrogen n_H (units of $[cm^{-3}]$) along a given *sightline*, $\int n_H dl$.

Neutrinos: very low-mass elementary particles that are copiously produced in fusion and fission processes but interact only very weakly with ordinary matter.

Neutron: an uncharged fundamental particle usually found in the nucleus of atoms heavier than hydrogen. It has about the same mass as a proton.

Neutron star (NS): a dense stellar remnant, much like a *white dwarf* but supported by *degeneracy pressure* associated with tightly packed neutrons. Typical sizes of NSs are about 10 km in radius and about one Solar mass. They are thought to form during the collapse of massive stars as those stars explode as supernovae.

NFI: Narrow Field X-ray Instruments, on board the *BeppoSAX* satellite.

Nucleosynthesis: the process of making new elements (nuclei) from existing elements. Usually nucleosynthesis is associated with the creation of new elements via fusion (rather than fission) processes. Explosive nucleosynthesis is the process of making a large mass of new elements rapidly (on seconds timescales); this happens at the start of a supernova.

Occam's Razor: the notion that the simplest explanation is likely the correct one; a guiding principle in scientific pursuits.

Optical depth (τ): a dimensionless number related to the stopping power of light. High optical depth ($\tau \gg 1$) implies that that material is opaque to the transmission of light. Likewise, low optical depth ($\tau \ll 1$) material is transparent to the passage of light through it.

Pair production: the creation of an electron and positron, usually through the interaction (and annihilation) of two high-energy photons.

Panchromatic: across the electromagnetic spectrum, from radio to gamma-ray *wavebands*.

Parallax: the apparent wobble of a nearby object relative to more-distant objects on the celestial sphere, where a different vantage point is given to an observer on Earth due to the Earth's motion around the Sun. Parallax is used to find geometric distances to nearby stars.

Parsec: a measure of distance on astronomical scales equal to 3.085×10^{13} km. Distances to nearby stars are usually given in units of parsec. The word parsec is a conjunction of the words *parallax* and *arcsecond*, since it is the distance at which a source has an apparent parallax of one arcsecond.

Periodic: a regular and repeating change in brightness. There is no periodicity seen in GRBs, but SGR light curves are seen to change periodically.

Photometric redshift: the technique of using an afterglow imaged at different wavebands to infer the redshift of a source.

Photon: a parcel (or particle) of light that carries energy and momentum. Light can be considered both a wave and a particle. This is an utterly stupefying concept, but Nature has asked us to wrap our heads around this duality. When considering the detection of light from astronomical sources, it is usual to consider optical, X-ray, and gamma-ray light as individual photons. For long-wavelength light, such as radio light, it is usual to consider the energy carried as waves.

Plane wave: if we think of the propagation of light through space as the motion of the crests of waves (like the crests of ocean waves heading toward a beach), then the crest of a plane wave lies perpendicular to the direction of motion. On a placid lake where a pebble has just dropped, the crests are curved and form concentric rings that move outward. Far from where the pebble has dropped (at later times), the crests will nearly appear as plane waves as they sweep by an observer. Likewise, when far from a light source, there is no apparent curvature in the crests.

Potential energy: the energy associated with a mass far from the local gravitational center.

Quasar: a light source that appears to be as compact as a star in the Milky Way but instead is a massive *black hole* in a distant galaxy. Quasars are thought to be powered by mass flowing into the BH and are typically tens of thousands of times brighter than the brightest galaxies.

Radian: a measure of angle. There are 2π radians in a circle of 360 degrees, so one radian is about 57.3 degrees.

Radiation: any form of energy, such as light or *gravitational waves*, that propagates through space.

Redshift: the apparent shift of a spectrum to longer wavelengths relative to what would be observed in a laboratory. Redshift is often a good-enough proxy for distance in astronomy: at cosmological distances, there is a one-to-one mapping between redshift of a source and its distance. Larger redshifts correspond to larger distances.

Relativistic expansion: outflow at velocities very near the speed of light.

Resolution (imaging):   a measure of the blurriness of an image, usually given in angular units ([radians] or [arcsec]) with smaller resolutions being more desirable.

Resolution (spectral):   a measure of the how spread out the light is in a spectrum. For analyzing the properties of gas clouds giving rise to afterglow absorption, higher-resolution spectra are generally more desirable.

Restmass energy: the energy equivalent to the mass M of the object, related by $E = Mc^2$.

SF: star formation.

SFR: star-formation rate, usually given in units of solar mass per year per unit volume.

Shock: a region in space where there is a sharp difference (i.e., a discontinuity) in density, temperature, and/or pressure.

Shockwave: a propagating *shock* in space and time. A supernova shockwave propagates radially from the center of the explosion.

Sightline: a direction toward or away from a specific astrophysical source through which light travels. Absorbing material in the sightline to that distant source will often leave a characteristic fingerprint in the observed spectrum of that source. See *absorption-line spectroscopy*.

SMC: Small Magellanic Cloud. See also the *LMC*.

Soft Gamma-ray Repeaters (SGRs):   a class of bursting objects that appear to repeatedly produce gamma-ray events over hours to decades (whereas classical GRBs do not repeat) and show *periodic* behavior during the gamma-ray event. SGRs are thought to come from *magnetars*.

Spectral-energy distribution (SED): the *spectrum* of a source over many decades in frequency or wavelength. A spectrum is the brightness (in flux or luminosity units usually) versus wavelength or frequency.

Spectrum: the brightness of a source as a function of wavelength (or frequency). As a prism splits a light beam into its constituent colors, modern instruments ("spectrographs") disperse the light of a GRB afterglow to reveal its spectrum.

Supernova (SN): the result of a violent explosion of a star (or two merged stars) whereby a significant amount of mass ($>0.1 M_\odot$) is expelled at large velocity (>thousand km/s). Supernovae are heated by radioactive elements created at the outset of the explosion. This heating, coupled with the expansion of the ejected material (*ejecta*), leads to a characteristic rising and falling behavior at optical wavelengths; the time-to-rise of most SNe is from days to weeks following the explosion.

Supernova remnant: the bright, dense, clumpy (and partially radioactive) material ejected during a supernova event. Since the *ejecta* can move at large velocities, supernova remnants can become very large and likewise appear to have a large angular size. The famous Crab nebula is the remnant of a supernova which was discovered in the year 1054 (see http://wikipedia.org/wiki/Crab_Nebula).

Synchrotron blastwave theory: a theoretical framework that describes the origin of the afterglow emission, connecting the bulk properties of outflowing material (such as velocity and mass), relativistic *shocks*, and *synchrotron radiation.*

Synchrotron radiation: light produced by charged particles (usually electrons) moving near the speed of light (i.e., relativistically) in a magnetic field.

Timescale: the characteristic length of time over which there are significant changes in some observed property (e.g., brightness/*flux*).

UVOIR: a span over the optical, infrared, and ultraviolet portions of the electromagnetic spectrum, from several hundreds of Ångström to a few ten thousands of Ångström.

UVOT: the Ultraviolet-Optical Telescope, on board Swift.

Vela: a series of U.S. satellites operating from the 1960s and 1970s.

VLA: the Very Large Array; a radio observatory located in New Mexico.

Waveband: a certain span in wavelength, frequency, or energy of the *electromagnetic spectrum.*

WFC: Wide-Field Camera, on board the BeppoSAX satellite.

White dwarf (WD): the remnant core of a former star that is no longer undergoing fusion processes. This dense and compact object is supported against gravitational collapse by *degeneracy* *pressure* associated with the tight packing of electrons. The brightness of a WD is determined by the rate at which residual heat leaks out from the surface of the object. The typical size of a WD is comparable to the size of the Earth but has a mass similar to that of the Sun.

$\bar{\chi}_H(z)$: the average neutral fractional of hydrogen at a given *redshift* (z). The value $\bar{\chi}_H(z)$ is the ratio of the number density of neutral hydrogen atoms divided by the number density of neutral hydrogen atoms plus protons.

X-ray: wavelength regime of the electromagnetic spectrum between ultraviolet and gamma ray; photon energies from 0.1 keV to $\sim$10 keV.

XRT: the X-ray Telescope, on board Swift.

INDEX